Encyclopaedia of

World Geography

Encyclopaedia of

World Geography

Vol. - 6

Chandra Vijay Purty

ABD PUBLISHERS

Jaipur ■ New Delhi

ABD PUBLISHERS

Regd. Off. "Bony Residency", Gate No. 2
Opp. Tilak Public School, Vishweswariya Nagar,
Gopalpura Road, Jaipur - 302018, India
Ph: 0141 - 2761280, Telefax: 2761381
Email: abdpublishersjpr@gmail.com

Delhi Office: 102, 1st Floor, Satyam House,
4327/3, Ansari Road, Darya Ganj,
New Delhi - 110002, India
Ph: 011-45652440

visit us: www.oxfordbookcompany.com

Published in India, 2024

ISBN : 9788183763127 (7 Vols. Set)

Typeset by Shivangi Computer, Jaipur

Printed and bound in India by Thomson Press Ltd.,
New Delhi

Preface

Geography is the study of the Earth and its lands, features, inhabitants, and phenomena. A literal translation would be "to describe or write about the Earth". The first person to use the word "geography" was Eratosthenes (276-194 B.C.). Four historical traditions in geographical research are the spatial analysis of natural and human phenomena (geography as a study of distribution), area studies (places and regions), study of man-land relationship, and research in earth sciences. Nonetheless, modern geography is an all-encompassing discipline that foremost seeks to understand the Earth and all of its human and natural complexities—not merely where objects are, but how they have changed and come to be. As "the bridge between the human and physical sciences," geography is divided into two main branches—human geography and physical geography.

The oldest known world maps date back to ancient Babylon from the 9th century BC. The best known Babylonian world map, however, is the *Imago Mundi* of 600 BC. The map as reconstructed by Eckhard Unger shows Babylon on the Euphrates, surrounded by a circular landmass showing Assyria, Urartu and several cities, in turn surrounded by a "bitter river" (Oceanus), with seven islands arranged around it so as to form a seven-pointed star. The accompanying text mentions seven outer regions beyond the encircling ocean. The descriptions of five of them have survived. In contrast to the *Imago Mundi*, an earlier Babylonian world map dating back to the 9th century BC depicted Babylon as being further north from the centre of the world, though it is not certain what that centre was supposed to represent.

The ideas of Anaximander (c. 610 B.C.-c. 545 B.C.), considered by later Greek writers to be the true founder of geography, come to us through fragments quoted by his successors. Anaximander is credited with the invention of the gnomon, the simple yet efficient Greek instrument that allowed the early measurement of latitude. Thales, Anaximander is also credited with the prediction of eclipses. The foundations of geography can be traced to the ancient cultures, such as the ancient, medieval, and early modern Chinese. The Greeks, who were the first to explore geography as both art and science, achieved this through Cartography, Philosophy, and Literature, or through Mathematics. There is some debate about who was the first person to assert that the Earth is spherical in shape, with the credit going either to Parmenides or Pythagoras. Anaxagoras was able to demonstrate that the profile of the Earth was circular by explaining eclipses. However, he still believed that the Earth was a flat disk, as did many of his contemporaries. One of the first estimates of the radius of the Earth was made by Eratosthenes.

The first rigorous system of latitude and longitude lines is credited to Hipparchus. He employed a sexagesimal system that was derived from Babylonian mathematics. The parallels and meridians were sub-divided into 360°, with each degree further subdivided 602 (minutes). To measure the longitude at different location on Earth, he suggested using eclipses to determine the relative difference in time. The extensive mapping by the Romans as they explored new lands would later provide a high level of information for Ptolemy to construct detailed atlases. He extended the work of Hipparchus, using a grid system on his maps and adopting a length of 56.5 miles for a degree.

From the 3rd century onwards, Chinese methods of geographical study and writing of geographical literature became much more complex than what was found in Europe at the time (until the 13th century). Chinese geographers such as Liu An, Pei Xiu, Jia Dan, Shen Kuo, Fan Chengda, Zhou Daguan, and Xu Xiake wrote important treatises, yet by the 17th century, advanced ideas and methods of Western-style geography were adopted in China.

This book contains the fundamental and basic information of the subject.

Chandra Vijay Purty

Contents

1

The Republic of Guinea

Guinea, is a country in West Africa. Formerly known as French Guinea (*Guinee francaise*), it is today sometimes called Guinea-Conakry to distinguish it from its neighbour Guinea-Bissau. Conakry is the capital, the seat of the national government, and the largest city.

Guinea has almost 246,000 square kilometres (94,981 sq mi). It forms a crescent by curving from its western border on the Atlantic Ocean toward the east and the south. Guinea shares its northern border with Guinea-Bissau, Senegal, and Mali. Guinea shares its southern border with Sierra Leone, Liberia, and Cote d'Ivoire. The Niger River arises in Guinea and runs eastward.

Guinea is home to twenty-four ethnic groups. The most prominent groups are the Fula, Mandinka, and Susu.

History

Guinea belonged to a series of empires until France colonized it in the 1890s, and made it part of French West Africa. Guinea declared its independence from France on 2 October 1958. Since its independence, Guinea has had autocratic rulers who have made Guinea one of the poorest countries.

Governments Since Independence

Ahmed Sekou Toure became President upon Guinea's independence. By violent repression, he ruled until 26 March 1984, when he died unexpectedly. By a quick coup d'etat, Lansana Conte became the President after Toure. By despotic means, Conte clung to power until his death in 2008. Despite extraordinary aluminium rich resources, he was unable to improve the desperate economic plight into which Toure had plunged the country.

On 23 December 2008, Moussa Dadis Camara seized control of Guinea as the head of a junta. On 28 September 2009, the junta ordered its soldiers to attack people who had gathered to protest any attempt by Camara to become President. The soldiers went on a rampage of rape, mutilation, and murder.

On 3 December 2009, an aide shot Camara during a dispute about the rampage of September 2009. Camara went to Morocco for medical care. Vice-President (and defence minister) Sekouba Konnte flew back from Lebanon to run the country in Camara's absence.

On January 12, 2010 Camara was flown from Morocco to Burkina Faso. After meeting in Ouagadougou on January 13 and 14, Camara, Konnte and Blaise Compaore, President of Burkina Faso, produced a formal statement of twelve principles promising a return of Guinea to civilian rule within six months. It was agreed that the military would not contest the forthcoming elections, and Camara would continue his convalescence outside Guinea. On 21 January 2010 the military junta appointed Jean-Marie Dore as Prime Minister of a six-month transition government, leading up to elections.

The presidential election will be held on 27 June and 18 July 2010.

Regions and Prefectures

The Republic of Guinea covers 245,857 square kilometres (94,926 sq mi) of West Africa about 10 degrees north of the equator. Guinea is divided into four natural regions with distinct human, geographic, and climatic characteristics:

- Maritime Guinea (*La Guinee Maritime*) covers 18% of the country
- Mid-Guinea (*La Moyenne-Guinee*) covers 20% of the country
- Upper-Guinea (*La Haute-Guinee*) covers 38% of the country

Forested Guinea (*Guinee Forestiere*) covers 23% of the country, and is both forested and mountainous.

Guinea is divided into seven administrative regions and subdivided into thirty-three prefectures. The national capital, Conakry, ranks as a special zone.

Region	*Capital*	*Population*
Kankan Region	Kankan	540,427
Kindia Region	Kindia	520,807
Nzerekore Region	Nzerekore	660,400
Labe Region	Labe	480,740
Faranah Region	Faranah	440,500
Mamou Region	Mamou	434,600

The capital Conakry with a population of 2,782,630 ranks as a special zone.

Geography

At 94,919 square miles (245,839 km), Guinea is roughly the size of the United Kingdom and slightly smaller than the U.S. state of Oregon. There are 200 miles (322 km) of coastline and a total land border of 2,112 miles (3,399 km). Its neighbours are Cote d'Ivoire (Ivory Coast), Guinea-Bissau, Liberia, Mali, Senegal and Sierra Leone.

The country is divided into four main regions: the Basse-Cote lowlands, populated mainly by the Susu ethnic group; the cooler, mountainous Fouta Djallon that run roughly north-south through the middle of the country, populated by Peuls, the Sahelian Haute-Guinea to the northeast, populated by Malinke, and the forested jungle regions in the southeast, with several ethnic groups. Guinea's mountains are the source for the Niger, the Gambia, and Senegal Rivers, as well as the numerous rivers flowing to the sea on the west side of the range in Sierra Leone and Ivory Coast.

Economy

Guinea has abundant natural resources including 25% or more of the world's known bauxite reserves. Guinea also has diamonds, gold, and other metals. The country has great potential for hydroelectric power. Bauxite and alumina are currently the only major exports. Other industries include processing plants for beer, juices, soft drinks and tobacco. Agriculture employs 80% of the nation's labour force. Under French rule, and at the beginning of independence, Guinea was a major exporter of bananas, pineapples, coffee, peanuts, and palm oil.

Mining

Richly endowed with minerals, Guinea possesses over 25 billion tonnes (metric tons) of bauxite – and perhaps up to one-half of the world's reserves. In addition, Guinea's mineral wealth includes more than 4-billion tonnes of high-grade iron ore, significant diamond and gold deposits, and undetermined quantities of uranium. Guinea has considerable potential for growth in agricultural and fishing sectors. Soil, water, and climatic conditions provide opportunities for large-scale irrigated farming and agro industry. Possibilities for investment and commercial activities exist in all these areas, but Guinea's poorly developed infrastructure and rampant corruption continue to present obstacles to large-scale investment projects.

Joint venture bauxite mining and alumina operations in northwest Guinea historically provide about 80% of Guinea's foreign exchange. The *Compagnie des Bauxites de Guinea* (CBG) is the main player in the bauxite industry. CBG is a joint venture, 49% owned by the Guinean Government and 51% by an international consortium led by Alcoa and Alcan. CBG exports about 14 million tonnes of high-grade bauxite annually. The *Compagnie des Bauxites de Kindia* (CBK), a joint venture between the Government of Guinea and Russki Alumina, produces some 2.5 million tonnes annually, nearly all of which is exported to Russia and Eastern Europe. Dian Dian, a Guinean/ Ukrainian joint bauxite venture, has a projected production rate of 1,000,000 per year, but is not expected to begin operations for several years. The *Alumina Compagnie de Guinee* (ACG), which took over the former Friguia Consortium, produced about 2.4 million tonnes in 2004 as raw material for its alumina refinery. The refinery exports about 750,000 tonnes of alumina. Both Global Alumina and Alcoa-Alcan have signed conventions with the Government of Guinea to build large alumina refineries with a combined capacity of about 4 million tonnes per year.

Diamonds and gold also are mined and exported on a large scale. AREDOR, a joint diamond-mining venture between the Guinean Government (50%) and an Australian, British, and Swiss consortium, began production in 1984 and mined diamonds that are 90% gem quality. Production stopped from 1993 until 1996, when First City Mining of Canada purchased the international portion of the consortium. The bulk of diamonds are mined artisanally. The largest gold mining operation in Guinea is a joint venture between the government and Ashanti Gold Fields of Ghana. SMD also has a large gold mining facility in Lero near the Malian border. Other concession agreements have been signed for iron ore, but these projects await preliminary exploration and financing results.

Problems and Reforms

The Guinean Government adopted policies in the 1990s to return commercial activity to the private sector, promote investment, reduce the role of the state in the economy, and improve the administrative and judicial framework. Guinea has the potential to develop, if the government carries out its announced policy reforms, and if the private sector responds appropriately. So far, corruption and favouritism, lack of long-term political stability, and lack of a transparent budgeting process continue to dampen foreign investor interest in major projects in Guinea.

Reforms since 1985 include eliminating restrictions on agriculture and foreign trade, liquidation of some parastatals, the creation of a realistic exchange rate, increased spending on education, and cutting the government bureaucracy. In July 1996, President Lansana Conte appointed a new government, which promised major economic reforms, including financial and judicial reform, rationalization of public expenditures, and improved government revenue collection. Under 1996 and 1998 International Monetary Fund (IMF)/World Bank agreements, Guinea continued fiscal reforms and privatization, and shifted governmental expenditures and internal reforms to the education, health, infrastructure, banking, and justice sectors.

The government revised the private investment code in 1998 to stimulate economic activity in the spirit of free enterprise. The code does not discriminate between foreigners and nationals and allows for repatriation of profits. While the code restricts development of Guinea's hydraulic resources to projects in which Guineans have majority shareholdings and management control, it does contain a clause permitting negotiations of more favourable conditions for investors in specific agreements. Foreign investments outside Conakry are entitled to more favourable terms. A national investment commission has been formed to review all investment proposals. Guinea and the United States have an investment guarantee agreement that offers political risk insurance to American investors through the Overseas Private Investment Corporation (OPIC). In addition, Guinea has inaugurated an arbitration court system, which allows for the quick resolution of commercial disputes.

Cabinet changes in 1999, which increased corruption, economic mismanagement, and excessive government spending, combined to slow the momentum for economic reform. The informal sector continues to be a major contributor to the economy.

Until June 2001, private operators managed the production, distribution, and fee-collection operations of water and electricity under performance-based contracts with the Government of Guinea. However, the two utilities are plagued by inefficiency and corruption. Foreign private investors in these operations departed the country in frustration.

In 2002, the IMF suspended Guinea's Poverty Reduction and Growth Facility (PRGF) because the government failed to meet key performance criteria. In reviews of the PRGF, the World Bank noted that Guinea had met its spending goals in targeted social priority sectors. However, spending in other areas, primarily defence, contributed to a significant fiscal deficit. The loss of IMF funds forced the government to finance its debts through Central Bank advances. The pursuit of unsound economic policies has resulted in imbalances that are proving hard to correct.

Under then-Prime Minister Diallo, the government began a rigorous reform agenda in December 2004 designed to return Guinea to a PRGF with the IMF. Exchange rates have been allowed to float, price controls on gasoline have been loosened, and government spending has been reduced while tax collection has been improved. These reforms have not reduced inflation, which hit 27% in 2004 and 30% in 2005. Currency depreciation is also a concern. The Guinea franc was trading at 2550 to the dollar in January 2005. It hit 5554 to the dollar by October 2006.

Despite the opening in 2005 of a new road connecting Guinea and Mali, most major roadways remain in poor repair, slowing the delivery of goods to local markets. Electricity and water shortages are frequent and sustained, and many businesses are forced to use expensive power generators and fuel to stay open.

Even though there are many problems plaguing Guinea's economy, not all foreign investors are reluctant to come to Guinea. Global Alumina's proposed alumina refinery has a price tag above $2 billion. Alcoa and Alcan are proposing a slightly smaller refinery worth about $1.5 billion. Taken together, they represent the largest private investment in sub-Saharan Africa since the Chad-Cameroon oil pipeline. Also, Hyperdynamics Corporation, an American oil company, signed an agreement in 2006 to develop Guinea's offshore Senegal Basin oil deposits in a 31,000 square mile concession; it is pursuing seismic exploration.

On 13 October 2009, Guinean Mines Minister Mahmoud Thiam announced that the Chinese International Fund would invest more than $7bn (£4.5bn) in infrastructure. In return, he said the firm would be a "strategic partner" in all mining projects in the mineral-rich nation. He said the firm would help build ports, railway lines, power plants, low-cost housing and even a new administrative centre in the capital, Conakry. However, analysts say that the timing of the deal is likely to stir controversy, as the legitimacy of Guinea's government is under question.

Youth unemployment, however, remains a large problem. Guinea needs an adequate policy to address the concerns of the urban youth. The problem is the disparity between their life and what they see on television. As the youth cannot find jobs, seeing the economically stable and consumerism of richer countries only serves to frustrate them further. Explosive youth

Oil

Northwest Africa's coast is now ripe for oil development, and Guinea is actively being courted in this endeavour. Guinea signed a Production sharing agreement with Hyperdynamics Corporation (Houston, TX) in 2006 to explore a large offshore tract, recently in partnership with Dana Petroleum PLC (Aberdeen, United Kingdom). Many petroleum experts believe that the offshore region from Ghana west to Senegal, which includes Guinea, might one day supply the United States nearly 30% of its oil needs before 2020.

Transportation

The railway which operated from Conakry to Kankan ceased operating in the mid-1980s. Domestic air services are intermittent. Most vehicles in Guinea are 20+ years old, and cabs are any four-door vehicle which the owner has designated as being for hire. Locals, nearly entirely without vehicles of their own, rely upon these taxis (which charge per seat) and small buses to take them around town and across the country. There is some river traffic on the Niger and Milo rivers. Horses and donkeys pull carts, primarily to transport construction materials.

Iron mining at Simandou in the southeast beginning in 2007 and at Kalia in the east is likely to result in the construction of a new heavy-duty standard gauge railway and deepwater port.

Demography

The population of Guinea is estimated at 10.2 million. Conakry, the capital and largest city, is the hub of Guinea's economy, commerce, education, and culture.

Languages

The official language of Guinea is French. Other significant languages spoken are Maninka (Malinke), Susu, Pular (Fulfulde or Fulani), Kissi, Kpelle, and Loma.

Ethnicity

The population of Guinea comprises about 24 ethnic groups. The Fulani, also known as Fula or Peul or Peuhl, comprise 40%

of the population and are mostly found in the Futa Jallon region. The Mandinka, also known as Mandingo or Malinke, comprise 30% of the population and are mostly found in eastern Guinea concentrated around the Kankan and Kissidougou prefectures. The Soussou, comprising 20%, are predominantly in western areas around the capital Conakry, Forecariah, and Kindia. Smaller ethnic groups make up the remaining 10% of the population, including Kpelle, Kissi, Zialo, Toma and others. Non-Africans total about 10,000 (mostly Lebanese, French, and other Europeans).

Religion

Islam is the majority religion. Approximately 85% of the population is Muslim, while 10% is Christian, and 5% holds traditional animist beliefs. Muslims are generally Sunni and Sufi; there are relatively few Shia. Christian groups include Roman Catholics, Anglicans, Baptists, Seventh-day Adventists, and other Evangelical groups. Jehovah's Witnesses are active in the country and recognized by the Government. There is a small Bahai community. There are small numbers of Hindus, Buddhists, and traditional Chinese religious groups among the expatriate community.

Military

The Guinean armed forces are divided into four branches. By far the largest branch of the armed forces, with about 15,000 personnel, the army is mainly responsible for protecting the state borders, the security of administered territories, and defending Guinea's national interests. Air force personnel total about 700. The force's equipment includes several Russian-supplied fighter planes and transports. The navy has about 900 personnel and operates several small patrol craft and barges. A branch of the Guinean Armed Forces is responsible for internal security. Its members are not police officers.

Health-care

Guinea has been reorganizing its health system since the Bamako Initiative of 1987 formally promoted community-based methods of increasing accessibility of drugs and health care services to the population, in part by implementing user fees.

The new strategy dramatically increased accessibility through community-based health-care (including community ownership and local budgeting), resulting in more efficient and equitable provision of services. A comprehensive strategy was extended to all areas of health care, with subsequent improvement in health indicators and improvement in health care efficiency and cost. Guinea's public health code is defined by Law No. L/97/021/AN of 19 June 1997 promulgating the Public Health Code. The law provides for the protection and promotion of health and for the rights and duties of the individual, the family, and community throughout the territory of the Republic of Guinea.

HIV/AIDS in Guinea

The first cases of HIV/AIDS were reported in 1986. Though levels of AIDS are significantly lower than in a number of other African countries, as of 2005, Guinea was considered by the World Health Organization to face a generalized epidemic.

An estimated 170,000 adults and children were infected at the end of 2004. The spread of the epidemic was attributed to factors such as proximity to high-prevalence countries, a large refugee population, internal displacement and subregional instability.

Culture

Like other West African countries, Guinea has a rich musical tradition. The group Bembeya Jazz became popular in the 1960s after Guinean independence.

2

The Republic of Ivory Coast

The Republic of Cote d'Ivoire, commonly known in English as Ivory Coast, is a country in West Africa. It has an area of 322,462 km^2, and borders the countries of Liberia, Guinea, Mali, Burkina Faso and Ghana; its southern boundary is along the Gulf of Guinea. The country's population which was 15,366,672 in 1998, was estimated to be 20,617,068 in 2009.

Prior to its occupation by Europeans, Cote d'Ivoire was home to several important states, including Gyaaman, the Kong Empire, and Baoule. There were also two Anyi kingdoms, *Indenie* and *Sanwi*, which attempted to retain their separate identity through the French colonial period and even after Cote d'Ivoire's independence. An 1843–1844 treaty made Cote d'Ivoire a "protectorate" of France and in 1893, it became a French colony as part of the European scramble for Africa.

Cote d'Ivoire became independent on 7 August 1960. From 1960 to 1993, the country was led by Felix Houphouët-Boigny. It maintained close political and economic association with its West African neighbours, while at the same time maintaining close ties to the West, especially to France. However, since the end of Houphouët-Boigny's rule, Cote d'Ivoire has experienced two coups d'etat (1999 and 2001) and a civil war, although elections and a political agreement between the new government and the rebels have brought a return to peace. Cote d'Ivoire is a republic with a strong executive power personified in the President. Its *de jure* capital is Yamoussoukro and the biggest city is the port city of Abidjan. The country is divided into 19 regions and 81 departments. It is a member of the Organisation of the Islamic Conference, African Union, La Francophonie, Latin Union, Economic Community of West African States and South Atlantic Peace and Cooperation Zone.

The official language is French although there are many other local languages, including Baoule, Dioula, Dan, Anyin and Cebaara Senufo. The main religions are Islam, Christianity (primarily Roman Catholic) and various indigenous religions.

The country, through its production of coffee and cocoa, was an economic powerhouse during the 1960s and 1970s in West Africa. However, Cote d'Ivoire went through an economic crisis in the 1980s, leading to the country's period of political and social turmoil. The 21st century Ivorian economy is largely market-based and relies heavily on agriculture, with smallholder cash crop production being dominant.

History

Land Migration

Of the first human presence in Cote d'Ivoire has been difficult to determine because human remains have not been well-preserved in the country's humid climate. However, the presence of new weapon and tool fragments (specifically, polished axes cut through shale and remnants of cooking and fishing) in the country has been interpreted as a possible indication of a large human presence during the Upper Paleolithic period (15,000 to 10,000 BC), or at the minimum, the Neolithic period.

The earliest known inhabitants of Cote d'Ivoire, however, have left traces scattered throughout the territory. Historians believe that they were all either displaced or absorbed by the ancestors of the present inhabitants, who arrived before the 16th century, including the Ehotile (Aboisso), Kotrowou (Fresco), Zehiri (Grand Lahou), Ega and Dies (Divo).

Pre-Islamic and Islamic Periods

The first recorded history is found in the chronicles of North African (Berber) traders, who, from early Roman times, conducted a caravan trade across the Sahara in salt, slaves, gold, and other goods. The southern terminals of the trans-Saharan trade routes were located on the edge of the desert, and from there supplemental trade extended as far south as the edge of the rain forest. The more important terminals – Djenne, Gao, and Timbuctu – grew into major commercial centres around which the great Sudanic empires developed.

By controlling the trade routes with their powerful military forces, these empires were able to dominate neighbouring states. The Sudanic empires also became centres of Islamic education. Islam had been introduced into the western Sudan (today's Mali) by Muslim Berber traders from North Africa and spread rapidly after the conversion of many important rulers. From the eleventh century, by which time the rulers of the Sudanic empires had embraced Islam, it spread south into the northern areas of contemporary Cote d'Ivoire.

The Ghana empire, the earliest of the Sudanic empires, flourished in present-day eastern Mauretania from the fourth to the thirteenth century. At the peak of its power in the eleventh century, its realms extended from the Atlantic Ocean to Timbuctu. After the decline of Ghana, the Mali Empire grew into a powerful Muslim state, which reached its apogee in the early part of the fourteenth century. The territory of the Mali Empire in Cote d'Ivoire was limited to the northwest corner around Odienne.

Its slow decline starting at the end of the fourteenth century followed internal discord and revolts by vassal states, one of which, Songhai, flourished as an empire between the fourteenth and sixteenth centuries. Songhai was also weakened by internal discord, which led to factional warfare. This discord spurred most of the migrations of peoples southward toward the forest belt. The dense rain forest covering the southern half of the country created barriers to large-scale political organizations as seen further north. Inhabitants lived in villages or clusters of villages whose contacts with the outside world were filtered through long-distance traders. Villagers subsisted on agriculture and hunting.

Five important states flourished in Cote d'Ivoire in the pre-European era. The Muslim Kong Empire was established by the Juula in the early eighteenth century in the north-central region inhabited by the Senoufo, who had fled Islamization under the Mali Empire. Although Kong became a prosperous centre of agriculture, trade, and crafts, ethnic diversity and religious discord gradually weakened the kingdom. The city of Kong was destroyed in 1895 by Samori Ture.

The Abron kingdom of Gyaaman was established in the seventeenth century by an Akan group, the Abron, who had fled the developing Ashanti confederation of Asanteman in what is present-day Ghana. From their settlement south of Bondoukou, the Abron gradually extended their hegemony over the Dyula people in Bondoukou, who were recent emigres from the market city of Begho. Bondoukou developed into a major centre of commerce and Islam. The kingdom's Quranic scholars attracted students from all parts of West Africa. In the mid-Seventeen century in east-central Cote d'Ivoire, other Akan groups fleeing the Asante established a Baoule kingdom at Sakasso and two Agni kingdoms, Indenie and Sanwi.

The Baoule, like the Ashanti, elaborated a highly centralized political and administrative structure under three successive rulers, but it finally split into smaller chiefdoms. Despite the breakup of their kingdom, the Baoule strongly resisted French subjugation. The descendants of the rulers of the Agni kingdoms tried to retain their separate identity long after Cote d'Ivoire's independence; as late as 1969, the Sanwi of Krinjabo attempted to break away from Cote d'Ivoire and form an independent kingdom.

Establishment of French Rule

Compared to neighbouring Ghana, Cote d'Ivoire suffered little from the slave trade, as European slaving and merchant ships preferred other areas along the coast, with better harbors. The earliest recorded French voyage to West Africa took place in 1483. The first West African French settlement, Saint Louis, was founded in the mid-seventeenth century in Senegal, while at about the same time the Dutch ceded to the French a settlement at Goree Island off Dakar. A French mission was established in 1637 Assinie near the border with the Gold Coast (now Ghana).

Assinie's survival was precarious, however, and only in the mid-nineteenth century did the French establish themselves firmly in Cote d'Ivoire. In 1843–1844, French admiral Bouet-

Willaumez signed treaties with the kings of the Grand Bassam and Assinie regions, placing their territories under a French protectorate. French explorers, missionaries, trading companies, and soldiers gradually extended the area under French control inland from the lagoon region. However, pacification was not accomplished until 1915.

Activity along the coast stimulated European interest in the interior, especially along the two great rivers, the Senegal River and the Niger River. Concerted French exploration of West Africa began in the mid-nineteenth century but moved slowly and was based more on individual initiative than on government policy. In the 1840s, the French concluded a series of treaties with local West African rulers that enabled the French to build fortified posts along the Gulf of Guinea to serve as permanent trading centres.

The first posts in Cote d'Ivoire included one at Assinie and another at Grand Bassam, which became the colony's first capital. The treaties provided for French sovereignty within the posts and for trading privileges in exchange for fees or *coutumes* paid annually to the local rulers for the use of the land. The arrangement was not entirely satisfactory to the French because trade was limited and misunderstandings over treaty obligations often arose. Nevertheless, the French government maintained the treaties, hoping to expand trade.

France also wanted to maintain a presence in the region to stem the increasing influence of the British along the Gulf of Guinea coast. Thereafter, the French built naval bases to keep out non-French traders and began a systematic conquest of the interior. (They accomplished this only after a long war in the 1890s against Mandinka forces, mostly from Gambia. Guerrilla warfare by the Baoule and other eastern groups continued until 1917.)

The defeat of France in the Franco-Prussian War in 1871 and the subsequent annexation by Germany of the French province of Alsace Lorraine caused the French government to abandon its colonial ambitions and withdraw its military garrisons from its French West African trading posts, leaving them in the care of resident merchants. The trading post at Grand Bassam in Cote d'Ivoire was left in the care of a shipper from Marseille, Arthur Verdier, who in 1878 was named Resident of the Establishment of Cote d'Ivoire.

In 1886, to support its claims of effective occupation, France again assumed direct control of its West African coastal trading posts and embarked on an accelerated program of exploration in the interior. In 1887 Lieutenant Louis Gustave Binger began a two-year journey that traversed parts of Cote d'Ivoire's interior. By the end of the journey, he had concluded four treaties establishing French protectorates in Cote d'Ivoire. Also in 1887, Verdier's agent, Marcel Treich-Laplene, negotiated five additional agreements that extended French influence from the headwaters of the Niger River Basin through Cote d'Ivoire.

French Colonial Era

By the end of the 1880s, France had established what passed for effective control over the coastal regions of Cote d'Ivoire, and in 1889 Britain recognized French sovereignty in the area. That same year, France named Treich-Laplene titular governor of the territory. In 1893 Cote d'Ivoire was made a French colony, and then Captain Binger was appointed governor. Agreements with Liberia in 1892 and with Britain in 1893 determined the eastern and western boundaries of the colony, but the northern boundary was not fixed until 1947 because of efforts by the French government to attach parts of Upper Volta (present-day Burkina Faso) and French Sudan (present-day Mali) to Cote d'Ivoire for economic and administrative reasons.

France's main goal was to stimulate the production of exports. Coffee, cocoa and palm oil crops were soon planted along the coast. Cote d'Ivoire stood out as the only West African country with a sizeable population of settlers; elsewhere in West and Central Africa, the French and British were largely bureaucrats. As a result, a third of the cocoa, coffee and banana plantations were in the hands of French citizens and a forced-labour system became the backbone of the economy.

Throughout the early years of French rule, French military contingents were sent inland to establish new posts. The African population resisted French penetration and settlement. Among those offering greatest resistance was Samori Ture, who in the

1880s and 1890s was establishing the Wassoulou Empire which extended over large parts of present-day Guinea, Mali, Burkina Faso, and Cote d'Ivoire. Samori Ture's large, well-equipped army, which could manufacture and repair its own firearms, attracted strong support throughout the region. The French responded to Samori Ture's expansion of regional control with military pressure. French campaigns against Samori Ture, which were met with fierce resistance, intensified in the mid-1890s until he was captured in 1898.

France's imposition of a head tax in 1900, aimed at enabling the colony to undertake a public works program, provoked a number of revolts. Ivoirians viewed the tax as a violation of the terms of the protectorate treaties because it seemed that France was now demanding the equivalent of a *coutume* from the local kings rather than the reverse. Much of the population, especially in the interior, also considered the tax a humiliating symbol of submission.

From 1904 to 1958, Cote d'Ivoire was a constituent unit of the Federation of French West Africa. It was a colony and an overseas territory under the Third Republic. Until the period following World War II, governmental affairs in French West Africa were administered from Paris. France's policy in West Africa was reflected mainly in its philosophy of "association", meaning that all Africans in Cote d'Ivoire were officially French "subjects" without rights to representation in Africa or France.

French colonial policy incorporated concepts of assimilation and association. Assimilation presupposed the inherent superiority of French culture over all others, so that in practice the assimilation policy in the colonies meant extension of the French language, institutions, laws, and customs. The policy of association also affirmed the superiority of the French in the colonies, but it entailed different institutions and systems of laws for the colonizer and the colonized. Under this policy, the Africans in Cote d'Ivoire were allowed to preserve their own customs insofar as they were compatible with French interests.

An indigenous elite trained in French administrative practice formed an intermediary group between the French and the Africans. Assimilation was practiced in Cote d'Ivoire to the extent that after 1930 a small number of Westernized

Ivoirians were granted the right to apply for French citizenship. Most Ivoirians, however, were classified as French subjects and were governed under the principle of association. As subjects of France they had no political rights. Moreover, they were drafted for work in mines, on plantations, as porters, and on public projects as part of their tax responsibility. They were also expected to serve in the military and were subject to the indigenat, a separate system of law.

In World War II, the Vichy regime remained in control until 1943, when members of Gen. Charles de Gaulle's provisional government assumed control of all French West Africa. The Brazzaville Conference of 1944, the first Constituent Assembly of the Fourth Republic in 1946, and France's gratitude for African loyalty during World War II led to far-reaching governmental reforms in 1946. French citizenship was granted to all African "subjects," the right to organize politically was recognized, and various forms of forced labour were abolished.

Until 1958, governors appointed in Paris administered the colony of Cote d'Ivoire, using a system of direct, centralized administration that left little room for Ivoirian participation in policy making. The English colonial administration also adopted divide-and-rule policies, applying ideas of assimilation only to the educated elite. The French were also interested in ensuring that the small but influential elite was sufficiently satisfied with the status quo to refrain from any anti-French sentiment. In fact, although they were strongly opposed to the practices of association, educated Ivoirians believed that they would achieve equality with their French peers through assimilation rather than through complete independence from France, a change that would eliminate the enormous economic advantages of remaining a French possession. But after the assimilation doctrine was implemented entirely, at least in principle, through the postwar reforms, Ivoirian leaders realized that even assimilation implied the superiority of the French over the Ivoirians and that discrimination and inequality would end only with independence.

Independence

The son of a Baoule chief, Felix Houphouët-Boigny, was to become Cote d'Ivoire's father of independence. In 1944 he formed the country's first agricultural trade union for African cocoa farmers like himself. Angered that colonial policy favoured French plantation owners, they united to recruit migrant workers for their own farms. Houphouët-Boigny soon rose to prominence and within a year was elected to the French Parliament in Paris. A year later the French abolished forced labour. Houphouët-Boigny established a strong relationship with the French government, expressing a belief that the country would benefit from it, which it did for many years. France made him the first African to become a minister in a European government.

A turning point in relations with France was reached with the 1956 Overseas Reform Act (*Loi Cadre*), which transferred a number of powers from Paris to elected territorial governments in French West Africa and also removed remaining voting inequalities. In 1958, Cote d'Ivoire became an autonomous member of the French Community (which replaced the French Union). At the time of Cote d'Ivoire's independence (1960), the country was easily French West Africa's most prosperous, contributing over 40% of the region's total exports. When Houphouët-Boigny became the first president, his government gave farmers good prices for their products to further stimulate production. Coffee production increased significantly, catapulting Cote d'Ivoire into third place in world output (behind Brazil and Colombia). By 1979 the country was the world's leading producer of cocoa.

It also became Africa's leading exporter of pineapples and palm oil. French technicians contributed to the 'Ivoirian miracle'. In the rest of Africa, Europeans were driven out following independence; but in Cote d'Ivoire, they poured in. The French community grew from only 30,000 prior to independence to 60,000 in 1980, most of them teachers, managers and advisors. For 20 years, the economy maintained an annual growth rate of nearly 10% – the highest of Africa's non-oil-exporting countries.

Houphouet-Boigny Administration

Politically, Houphouet-Boigny ruled with a firmness some called an "iron hand"; others characterized his rule more mildly as "paternal." The press was not free and only one political party existed, although some accepted this as a consequence of Houphouet-Boigny's broad appeal to the population that continually elected him. He was also criticized for his emphasis on developing large scale projects. Many felt the millions of dollars spent transforming his home village, Yamoussoukro, into the new capital that it became, were wasted; others support his vision to develop a centre for peace, education and religion in the heart of the country. But in the early 1980s, the world recession and a local drought sent shock waves through the Ivoirian economy. Due to the over cutting of timber and collapsing sugar prices, the country's external debt increased threefold. Crime rose dramatically in Abidjan.

In 1990, hundreds of civil servants went on strike, joined by students protesting institutional corruption. The unrest forced the government to support multi-party democracy. Houphouet-Boigny became increasingly feeble and died in 1993. He favoured Henri Konan Bedie as his successor.

Bedie Administration

In October 1995, Bedie overwhelmingly won re-election against a fragmented and disorganised opposition. He tightened his hold over political life, jailing several hundred opposition supporters. In contrast, the economic outlook improved, at least superficially, with decreasing inflation and an attempt to remove foreign debt.

Unlike Houphouet-Boigny, who was very careful in avoiding any ethnic conflict and left access to administrative positions wide-open to immigrants from neighbouring countries, Bedie emphasized the concept of "Ivority" (French: *Evirate*) to exclude his rival Alassane Ouattara, who had two northern Ivorian parents, from running for future presidential election. As people originating from foreign countries are a large part of the Ivoirian population, this policy excluded many people from Ivoirian nationality, and the relationship between various ethnic groups became strained.

1999 Coup

Similarly, Bedie excluded many potential opponents from the army. In late 1999, a group of dissatisfied officers staged a military coup, putting General Robert Guei in power. Bedie fled into exile in France. The new leadership reduced crime and corruption, and the generals pressed for austerity and openly campaigned in the streets for a less wasteful society.

Gbagbo Administration

A presidential election was held in October 2000 in which Laurent Gbagbo vied with Guei, but it was peaceful. The lead-up to the election was marked by military and civil unrest. Following a public uprising that resulted in around 180 deaths, Guei was swiftly replaced by Gbagbo. Alassane Ouattara was disqualified by the country's Supreme Court, due to his alleged Burkinabe nationality. The existing and later reformed constitution [under Guei] did not allow non-citizens to run for presidency. This sparked violent protests in which his supporters, mainly from the country's north, battled riot police in the capital, Yamoussoukro.

Ivorian Civil War

In the early hours of September 19, 2002, while the President was in Italy, there was an armed uprising. Troops who were to be demobilised mutinied, launching attacks in several cities. The battle for the main gendarmerie barracks in Abidjan lasted until mid-morning, but by lunchtime the government forces had secured the main city, Abidjan. They had lost control of the north of the country, and the rebel forces made their stronghold in the northern city of Bouake. The rebels threatened to move on Abidjan again and France deployed troops from its base in the country to stop any rebel advance. The French said they were protecting their own citizens from danger, but their deployment also aided the government forces. It was not established as a fact that the French were helping either side but each side accused them of being on the opposite side. It is disputed as to whether the French actions improved or worsened the situation in the long term.

What exactly happened that night is disputed. The government said that former president Robert Guei had led a coup attempt, and state TV showed pictures of his dead body in the street; counter-claims said that he and fifteen others had been murdered at his home and his body had been moved to the streets to incriminate him. Alassane Ouattara took refuge in the French embassy, his home burned down. President Gbagbo cut short a trip to Italy and on his return stated, in a television address, that some of the rebels were hiding in the shanty towns where foreign migrant workers lived. Gendarmes and vigilantes bulldozed and burned homes by the thousands, attacking the residents.

An early ceasefire with the rebels, who had the backing of much of the northern populace, proved short-lived, and fighting over the prime cocoa-growing areas resumed. France sent in troops to maintain the cease-fire boundaries, and militias, including warlords and fighters from Liberia and Sierra Leone, took advantage of the crisis to seize parts of the west.

2002 Unity Government

In January 2003, Gbagbo and rebel leaders signed accords creating a "government of national unity". Curfews were lifted and French troops patrolled the western border of the country. Since then, the unity government has proven extremely unstable and the central problems remain with neither side achieving its goals. In March 2004, 120 people were killed in an opposition rally, and subsequent mob violence led to foreign nationals being evacuated. A later report concluded the killings were planned.

Though UN peacekeepers were deployed to maintain a *Zone of Confidence*, relations between Gbagbo and the opposition continued to deteriorate.

Early in November 2004, after the peace agreement had effectively collapsed following the rebels' refusal to disarm, Gbagbo ordered airstrikes against the rebels. During one of these airstrikes in Bouake, on November 6, 2004, French soldiers were hit and nine of them were killed; the Ivorian government has said it was a mistake, but the French have claimed it was deliberate. They responded by destroying most Ivoirian military aircraft (2 Su-25 planes and 5 helicopters), and violent retaliatory riots against the French broke out in Abidjan.

Gbagbo's original mandate as president expired on October 30, 2005, but due to the lack of disarmament it was deemed impossible to hold an election, and therefore his term in office was extended for a maximum of one year, according to a plan worked out by the African Union; this plan was endorsed by the United Nations Security Council. With the late October deadline approaching in 2006, it was regarded as very unlikely that the election would be held by that point, and the opposition and the rebels rejected the possibility of another term extension for Gbagbo. The U. N. Security Council endorsed another one-year extension of Gbagbo's term on November 1, 2006; however, the resolution provided for the strengthening of Prime Minister Charles Konan Banny's powers. Gbagbo said the next day that elements of the resolution deemed to be constitutional violations would not be applied.

A peace deal between the government and the rebels, or New Forces, was signed on March 4, 2007, and subsequently Guillaume Soro, leader of the New Forces, became prime minister. These events have been seen by some observers as substantially strengthening Gbagbo's position.

Regions and Departments

Cote d'Ivoire is divided into nineteen regions (*regions*):

1. Agneby
2. Bafing
3. Bas-Sassandra
4. Denguele
5. Dix-Huit Montagnes
6. Fromager
7. Haut-Sassandra
8. Lacs
9. Lagunes
10. Marahoue
11. Moyen-Cavally
12. Moyen-Comoe
13. N'zi-Comoe
14. Savanes
15. Sud-Bandama
16. Sud-Comoe
17. Vallee du Bandama

18. Worodougou
19. Zanzan.

The regions are further divided into 81 departments.

Population of Major Cities

The official capital of Cote d'Ivoire is Yamoussoukro (295,500), the fourth most populous city. Abidjan, with a population of 3,310,500, is the largest city and serves as the commercial and banking centre of Cote d'Ivoire as well as the de facto capital. It is also the most populous city in French-speaking Western Africa.

City	*Population*
Abidjan	3,310,500
Bouake	775,300
Daloa	489,100
Yamoussoukro	295,500
Korhogo	163,400
San Pedro	151,600
Divo	134,200

Politics

Since 1983, Cote d'Ivoire's official capital has been Yamoussoukro; Abidjan, however, remains the administrative centre. Most countries maintain their embassies in Abidjan, although some (including the United Kingdom) have closed their missions because of the continuing violence and attacks on Europeans. The Ivoirian population continues to suffer because of an ongoing civil war. International human rights organizations have noted problems with the treatment of captive non-combatants by both sides and the re-emergence of child slavery among workers in cocoa production.

Although most of the fighting ended by late 2004, the country remained split in two, with the north controlled by the *New Forces* (FN). A new presidential election was expected to be held in October 2005, and an agreement was reached among the rival parties in March 2007 to proceed with this, but it has since then been postponed numerous times due to delays in its preparation.

Geography

Cote d'Ivoire (the Ivory Coast) is a sub-Saharan nation in southern West Africa located at 8 00°N, 5 00°W. The country is shaped like a square and borders the Gulf of Guinea in the north Atlantic Ocean to the south (515 km of coastline) and five other African nations on the other three sides, with a total of 3,110 km of borders: Liberia to the southwest (716 km), Guinea to the northwest (610 km), Mali to the north-northwest (532 km), Burkina Faso to the north-northeast, and Ghana to the east (668 km). In total, Cote d'Ivoire comprises 322,460 km^2, of which 318,000 km^2 is land and 4,460 km^2 is water, which makes the country slightly larger than the U.S. state of New Mexico, or about the size of Germany.

Maritime Claims

Cote d'Ivoire makes maritime claims of 200 nautical miles (370.4 km; 230.2 mi) as an exclusive economic zone, 12 nautical miles (22.2 km; 13.8 mi) of territorial sea, and a 200 nautical miles (370.4 km; 230.2 mi) continental shelf.

Terrain and Topography

Cote d'Ivoire's terrain can generally be described as a large plateau rising gradually from sea level in the south to almost 500 m elevation in the north. The nation's natural resources have made it into a comparatively prosperous nation in the African economy. The southeastern region of Cote d'Ivoire is marked by coastal inland lagoons that starts at the Ghanaian border and stretch 300 km (186 mi) along the eastern half of the coast. The southern region, especially the southwest, is covered with dense tropical moist forest. The Eastern Guinean forests extend from the Sassandra River across the south-central and southeast portion of Cote d'Ivoire and east into Ghana, while the Western Guinean lowland forests extend west from the Sassandra River into Liberia and southeastern Guinea. The mountains of Dix-Huit Montagnes region, in the west of the country near the border with Guinea and Liberia, are home to the Guinean montane forests. The Guinean forest-savanna mosaic belt extends across the middle of the country from east to west, and is the transition zone between the coastal forests and the interior savannas. The forest-savanna mosaic interlaces forest, savanna and grassland habitats. Northern Cote d'Ivoire is part of the West Sudanian savanna, a savanna-and-scrubland zone of lateritic or sandy soils, with vegetation decreasing from south to north. The terrain is mostly flat to undulating plains, with mountains in the northwest. The lowest elevation in Cote d'Ivoire is at sea level on the coasts. The highest elevation is Mount Nimba, at 1,752 metres (5,748 ft) in the far west of the country along the border with Guinea and Liberia.

Climate

The climate of Cote d'Ivoire is generally warm and humid, ranging from equatorial in the southern coasts to tropical in the middle and semiarid in the far north. There are three seasons: warm and dry (November to March), hot and dry (March to May), and hot and wet (June to October).

Crops and Natural Resources

Cote d'Ivoire's also has a large timber industry due to its large forest coverage. The nation's hardwood exports match that of Brazil. In recent years there has been much concern about the rapid rate of deforestation. Rainforests are being destroyed at a rate sometimes cited as the highest in the world. The only forest left completely untouched in Cote d'Ivoire is Tai National Park (*Parc National de Tai*), a 3,600 km^2 (1,390 sq mi) area in the country's far southwest that is home to over 150 endemic species and many other endangered species such as the Pygmy Hippopotamus and 11 species of monkeys.

Eight percent of the country is arable land. Cote d'Ivoire is the world's largest producer of cocoa, a major national cash crop. Other chief crops include coffee, bananas, and oil palms, which produce palm oil and kernels. Natural resources include petroleum, natural gas, diamonds, manganese, iron, cobalt, bauxite, copper, and hydropower. There are more foreigners which makes up 67% of the country population than Ivorians.

Natural Hazards

Natural hazards include the heavy surf and the lack of natural harbours on the coast; during the rainy season torrential flooding is a danger.

Economy

The maintenance of close ties to France since independence in 1960, diversification of agriculture for export, and encouragement of foreign investment, have been factors in the economic growth of Cote d'Ivoire. In recent years Cote d'Ivoire has been subject to greater competition and falling prices in the global marketplace for its primary agricultural crops: coffee and cocoa. That, compounded with high internal corruption, makes life difficult for the grower and those exporting into foreign markets.

Cote d'Ivoire is a member of the Organization for the Harmonization of Business Law in Africa (OHADA).

Religion

Islam and Christianity are the major religions, practiced by most of the country's population. The rest maintain indigenous beliefs, or hold no religious beliefs.

Demographics

77% of the population are considered Ivoirians. They represent several different peoples and language groups. An estimated 65 languages are spoken in the country. One of the most common is Dyula, which acts as a trade language as well as a language commonly spoken by the Muslim population.

French, the official language, is taught in schools and serves as a lingua franca in the country. The native born population is roughly split into three groups of Muslim, Christian (primarily Roman Catholic) and animist.

Since Cote d'Ivoire has established itself as one of the most successful West African nations, about 20% of the population (about 3.4 million) consists of workers from neighbouring Liberia, Burkina Faso and Guinea. Over two thirds of these migrant workers are Muslim.

4% of the population is of non-African ancestry. Many are French, Lebanese, Vietnamese and Spanish citizens, as well as Protestant missionaries from the United States and Canada. In November 2004, around 10,000 French and other foreign nationals evacuated Cote d'Ivoire due to attacks from pro-government youth militias. Aside from French nationals, there are native-born descendants of French settlers who arrived during the country's colonial period.

Health

Life expectancy at birth was 41 for males in 2004, for females it was 47. Infant mortality was 118 of 1000 live births. There are 12 physicians per 100,000 people. About a quarter of the population live below the international poverty line of US$1.25 a day.

Education

A large part of the adult population, in particular women, are illiterate. Many children between 6 and 10 years are not enrolled in school. The majority of students in secondary education are male. At the end of secondary education, students can sit the Baccalaureat examination. The country has universities, including the University of Cote d'Ivoire.

Culture

Music

Each of the ethnic groups in Cote d'Ivoire has their own music genres, most showing strong vocal polyphony. Talking drums are also common, especially among the Appollo, and polyrhythms, another African characteristic, are found throughout Cote d'Ivoire and are especially common in the southwest.

Popular music genres from Cote d'Ivoire include zolbazo, zouglou and coupe-decale.

Sport

So far, Cote d'Ivoire has won one silver medal in the Olympics, for men's 400 metres. So far, Cote d'Ivoire has been at two men's football world cups.

Name of the Country

The region, and then the country, was originally known in English as *Ivory Coast*. In October 1985, the government requested that the country be known in every language as Cote d'Ivoire without a hyphen between the two words.

Usage in English

Despite the Ivorian government's request, the Anglicised rendering "Ivory Coast" (sometimes "the Ivory Coast") is still frequently used in English.

The BBC usually uses "Ivory Coast" both in news reports and on its page about the country, *The Guardian* newspaper's Style Guide says: "Ivory Coast, not 'The Ivory Coast' or 'Cote D'Ivoire'; its nationals are Ivorians", ABC News, *The Times*, *The New York Times*, and the South African Broadcasting Corporation all use "Ivory Coast" either exclusively or predominantly.

Many governments use "Cote d'Ivoire" for diplomatic reasons. The English country name registered with the United Nations and used by ISO 3166 is "Cote d'Ivoire". Other organizations that use "Cote d'Ivoire" include the United States Department of State (which uses "Cote d'Ivoire" in formal documents, but uses "Ivory Coast" in many general references, speeches and briefing documents), FIFA and the IOC (referring to their national football and Olympic teams in international games and in official broadcasts), *The Economist* news magazine, *Encyclopaedia Britannica*, and *National Geographic Society*.

3

Kenya

The Republic of Kenya is a country in East Africa. Lying along the Indian Ocean, at the equator, Kenya is bordered by Ethiopia (north), Somalia (northeast), Tanzania (south), Uganda plus Lake Victoria (west), and Sudan (northwest). Kenya has numerous wildlife reserves, containing thousands of animal species. The capital city is Nairobi. Kenya's population of nearly 39 million is diverse-with more than 40 different cultures represented. The country is named after Mount Kenya, a significant landmark and the second among the highest mountain peaks of Africa.

Etymology

The word Kenya, pronounced, originates from the Kikuyu, Embu and Kamba names for Mount Kenya, "*Kirinyaga*", "*Kirinyaa*" and "*Kiinyaa*". The English meaning of the words is "God's resting place" in all the three languages due to shared traditional beliefs that the Mountain is a sacred and holy place. Pre-historic volcanic eruptions of Mount Kenya (now extinct) may resulted in its association with divinity and creation among the indigenous Kikuyu-related ethic groups who are the original native inhabitants of the vast agricultural land surrounding Mount Kenya. In the 19th Century, the German explorer, Ludwig Krapf, recorded the name as both *Kenia* and *Kegnia* believed by some to be a corruption of the Kamba version.

During the colonial period the name was pronounced. That pronunciation has been abandoned since independence in favour of the African version.

Geography and climate

At 580,367 km^2 (224,081 sq mi), Kenya is the world's forty-seventh largest country (after Madagascar). From the coast on the Indian Ocean the Low plains rise to central highlands. The highlands are bisected by the Great Rift Valley; a fertile plateau in the east. The Kenyan Highlands comprise one of the most successful agricultural production regions in Africa. The highlands are the site of the highest point in Kenya (and the second highest in Africa): Mount Kenya, which reaches 5,199 m (17,057 ft) and is also the site of glaciers. Climate varies from tropical along the coast to arid in the interior. Mount Kilimanjaro (5,895 m/19,341 ft) can be seen from Kenya to the South of the Tanzanian border.

Kenya has a tropical climate. It is hot and humid at the coast, temperate inland and very dry in the north and northeast parts of the country. There is however a lot of rain between March and May (the *long rains*), and moderate rain in October and November (the *short rains*). The temperature remains high throughout these months. The country receives a great deal of sunshine all the year round and summer clothes are worn throughout the year. However, it is usually cool at night and early in the morning. The long rain season occurs from April to June. The short rain season occurs from October to December. The rainfall is sometimes heavy and often falls in the afternoons and evenings. The hottest period is from February to March and coldest in July to August.

Kenya has considerable land area of wildlife habitat, including the Masai Mara, where Blue Wildebeest and other bovids participate in a large scale annual migration. Up to 250,000 blue wildebeest perish each year in the long and arduous movement to find forage in the dry season. The "Big Five" animals of Africa can also be found in Kenya: the lion, leopard, buffalo, rhinoceros and elephant. A significant population of other wild animals, reptiles and birds can be found in the national parks and game reserves in the country.

The annual animal migration-especially migration of the wildebeest-occurs between June and September with millions of animals taking part. It has been a popular event for filmmakers to capture.

History

Prehistory

Giant crocodile fossils have been discovered in Kenya, dating from the Mesozoic Era, over 200 million years ago. The fossils were found in an excavation conducted by a team from the University of Utah and the National Museums of Kenya in July–August 2004 at Lokitaung Gorge, near Lake Turkana.

Fossils found in East Africa suggest that primates roamed the area more than 20 million years ago. Recent finds near Kenya's Lake Turkana indicate that hominids such as *Homo habilis* (1.8 and 2.5 million years ago) and *Homo erectus* (1.8 million to 350 000 years ago) are possible direct ancestors of modern *Homo sapiens* and lived in Kenya during the Pleistocene epoch. In 1984 one particular discovery made at Lake Turkana by famous palaeoanthropologist Richard Leakey and Kamoya Kimeu was the skeleton of a Turkana boy belonging to *Homo erectus* from 1.6 million years ago. Previous research on early hominids is particularly identified with Mary Leakey and Louis Leakey, who were responsible for the preliminary archaeological research at Olorgesailie and Hyrax Hill. Later work at the former was undertaken by Glynn Isaac.

Pre-colonial History

Cushitic-speaking people, as termed by Schloezer, from northern Africa, moved into the area that is now Kenya beginning around 2000 BCE. Arab traders began frequenting the Kenya coast around the 1st century CE. Kenya's proximity to the Arabian Peninsula invited colonization, and Arab and Persian settlements sprouted along the coast by the 8th century. Some of the "Arabs", like in much of East Africa, were Afro-Arabs. During the first millennium CE, Nilotic and Bantu-speaking peoples moved into the region, and the latter now comprise three-quarters of Kenya's population.

The Kenyan coast had served host to communities of ironworkers and communities of subsistence farmers, hunters and fishers who supported the economy with agriculture, fishing, metal production and trade with foreign countries. Around the 6th or 9th century CE Kenya switched to a maritime-based economy and began to specialize in shipbuilding to travel south by sea to other port cities such as Kilwa Masoko and Shanga along the East African coast. Mombasa became the major port city of pre-colonial Kenya in the Middle Ages and was used to trade with other African port cities, Persia, Arab traders, Yemen and even India. Fifteenth-century Portuguese voyager Duarte Barbosa claimed, "[Mombasa] is a place of great traffic and has a good harbour in which there are always moored small craft of many kinds and also great ships, both of which are bound from Sofala and others which come from Cambay and Melinde and others which sail to the island of Zanzibar."

In the centuries preceding colonization, the Swahili coast of Kenya was part of the east African region which traded with the Arab world and India especially for ivory and slaves (the Ameru tribe is said to have originated from slaves escaping from Arab lands some time around the year 1700). Initially these traders came mainly from Arab states, but later many also came from Zanzibar (such as Tippu Tip). Close to 90% of the population on the Kenya coast was enslaved.

Swahili, a Bantu language with Arabic, Persian, and other Middle Eastern and South Asian loan words, later developed as a *lingua franca* for trade between the different peoples.

The Luo of Kenya descend from early agricultural and herding communities from western Kenya's early pre-colonial history. The Luo along with other tribes associated with the Nilotic language group, are known to have originated from the north of Kenya, probably the northern regions of modern Sudan. The Nilotes, as they are known, are an anthropological group that originated from the northeastern regions of Africa. They may have moved south because of the wars that characterized the growth of territories such as Kush and Egypt. In Kenya this group comprises the Luo, Kalenjin, the Turkana and the Maasai as the main groups. This is evidenced by the presence of similar dialects among certain tribes in modern day Sudan. These tribes include the Acoli and Lwo (not same as Luo) who occupy modern Darfur region.

There are also other tribes belonging to this group in Uganda and Tanzania. This is attributed mainly to the Luo's affinity to Lake Victoria, which they have stuck to throughout the three countries (Uganda, Tanzania and Kenya). In Uganda they are known to have established the Buganda Kingdom and the Toro Kingdom. The Luo in Kenya are known to have fought numerous wars with their neighbours, notably the Kalenjin, for control of the lake.

Throughout the centuries the Kenyan Coast has played host to many merchants and explorers. Among the cities that line the Kenyan coast is the City of Malindi. It has remained an important Swahili settlement since the 14th century and once reviled Mombasa for dominance in this part of East Africa. Malindi has traditionally been a friendly port city for foreign powers. In 1414, the Arab Sultan of Malindi initiated diplomatic relations with Ming Dynasty China during the voyages of the explorer Zheng He. Malindi authorities welcomed Portuguese explorer, Vasco da Gama, in 1498.

Colonial History

The colonial history of Kenya dates from the establishment of a German protectorate over the Sultan of Zanzibar's coastal possessions in 1885, followed by the arrival of the Imperial British East Africa Company in 1888. Incipient imperial rivalry was forestalled when Germany handed its coastal holdings to Britain in 1890. This followed the building of the Kenya–Uganda railway passing through the country. This was resisted by some tribes — notably the Nandi led by *Orkoiyot* Koitalel Arap Samoei for ten years from 1895 to 1905 — still the British eventually built the railway. The *Nandi* were the first tribe to be put in a native reserve to stop them from disrupting the building of the railway. During the railway construction era, there was a significant inflow of Indian peoples who provided the bulk of the skilled manpower required for construction.

While building the railroad through Tsavo, that a number of the Indian railway workers and local African labourers were attacked by two lions known as the Tsavo maneaters. They and most of their descendants later remained in Kenya and formed the core of several distinct Indian communities such as the Ismaili Muslim and Sikh communities.

At the outbreak of World War I in August 1914, the governors of British East Africa (as the Protectorate was generally known) and German East Africa agreed a truce in an attempt to keep the young colonies out of direct hostilities. However Lt Col Paul von Lettow-Vorbeck took command of the German military forces, determined to tie down as many British resources as possible. Completely cut off from Germany, von Lettow conducted an effective guerilla warfare campaign, living off the land, capturing British supplies, and remaining undefeated. He eventually surrendered in Zambia eleven days after the Armistice was signed in 1918. To chase von Lettow the British deployed Indian Army troops from India and then needed large numbers of porters to overcome the formidable logistics of transporting supplies far into the interior by foot. The Carrier Corps was formed and ultimately mobilised over 400,000 Africans, contributing to their long-term politicisation.

During the early part of the 20th century, the interior central highlands were settled by British and other European farmers, who became wealthy farming coffee and tea. By the 1930s, approximately 30,000 white settlers lived in the area and gained a political voice because of their contribution to the market economy. The area was already home to over a million members of the Kikuyu tribe, most of whom had no land claims in European terms, and lived as itinerant farmers. To protect their interests, the settlers banned the growing of coffee, introduced a hut tax, and the landless were granted less and less land in exchange for their labour. A massive exodus to the cities ensued as their ability to provide a living from the land dwindled.

From October 1952 to December 1959, Kenya was under a state of emergency arising from the Mau Mau rebellion against British rule. The governor requested and obtained British and African troops, including the King's African Rifles. The British began counter-insurgency operations; in May 1953 General Sir George Erskine took charge as commander-in-chief of the colony's armed forces in May 1953, with the personal backing of Winston Churchill.

The capture of Warihii Itote (aka General China) on 15 January 1954, and the subsequent interrogation led to a better understanding of the Mau Mau command structure. Operation Anvil opened on 24 April 1954, after weeks of planning by the army with the approval of the War Council. The operation effectively placed Nairobi under military siege, and the occupants were screened and the Mau Mau supporters moved to detention camps. The Home Guard formed the core of the government's strategy as it was composed of loyalist Africans, not foreign forces like the British Army and King's African Rifles. By the end of the emergency the Home Guard had killed 4686 Mau Mau, amounting to 42% of the total insurgents. The capture of Dedan Kimathi on 21 October 1956, in Nyeri signified the ultimate defeat of the Mau Mau and essentially ended the military offensive. During this period, substantial governmental changes to land tenure occurred, the most important of which was the Swynnerton Plan, which was used to both reward loyalists and punish Mau Mau.

Post-colonial History

The first direct elections for Africans to the Legislative Council took place in 1957. Despite British hopes of handing power to "moderate" African rivals, it was the Kenya African National Union (KANU) of Jomo Kenyatta that formed a government shortly before Kenya became independent on 12 December 1963. During the same year, the Kenyan army fought the Shifta War against ethnic Somalis who wanted Kenya's NFD joined with the Republic of Somalia. The Shiftas were defeated in 1967.

To discourage further invasions, Kenya signed a defence pact with Ethiopia in 1969, which is still in effect.

On 12 December 1964 the Republic of Kenya was proclaimed, and Mzee Jomo Kenyatta became Kenya's first president. At Kenyatta's death in 1978, Daniel arap Moi became President. Daniel arap Moi retained the Presidency, being unopposed in elections held in 1979, 1983 (snap elections) and 1988, all of which were held under the single party constitution. The 1983 elections were held a year early, and were a direct result of an abortive military coup attempt on 1 August 1982.

The abortive coup was masterminded by a lowly ranked Air Force serviceman, Senior Private Hezekiah Ochuka and was staged mainly by enlisted men in the Air Force. The attempt was quickly suppressed by Loyalist forces led by the Army, the General Service Unit (GSU) — a paramilitary wing of the police — and later the regular police, but not without civilian casualties. This event led to the disbanding of the entire Air Force and a large number of its former members were either dismissed or court-martialled. The election held in 1988 saw the advent of the *mlolongo* (queuing) system, where voters were supposed to line up behind their favoured candidates instead of a secret ballot. This was seen as the climax of a very undemocratic regime and it led to widespread agitation for constitutional reform. Several contentious clauses, including one that allowed for only one political party were changed in the following years. In democratic, multiparty elections in 1992 and 1997, Daniel arap Moi won re-election. In 2002, Moi was constitutionally barred from running, and Mwai Kibaki, running for the opposition coalition "National Rainbow Coalition"—NARC, was elected President. Anderson (2003) reports the elections were judged free and fair by local and international observers, and seemed to mark a turning point in Kenya's democratic evolution.

Politics

Kenya is a presidential representative democratic republic, whereby the President is both the head of state and head of government, and of a multi-party system. Executive power is exercised by the government. Legislative power is vested in both the government and the National Assembly. The Judiciary is independent of the executive and the legislature. However, there was growing concern especially during former president Daniel Arap Moi's tenure that the executive was increasingly meddling with the affairs of the judiciary.

Kenya has maintained remarkable stability despite changes in its political system and crises in neighbouring countries. A cross-party parliamentary reform initiative in the autumn of 1997 revised some oppressive laws inherited from the colonial era that had been used to limit freedom of speech and assembly. This improved public freedoms and contributed to generally credible national elections in December 1997.

In December 2002, Kenyans held democratic and open elections, most of which were judged free and fair by international observers. The 2002 elections marked an important turning point in Kenya's democratic evolution in that power was transferred peacefully from the Kenya African Union (KANU), which had ruled the country since independence to the National Rainbow Coalition (Narc), a coalition of political parties. Under the presidency of Mwai Kibaki, the new ruling coalition promised to focus its efforts on generating economic growth, combating corruption, improving education, and rewriting its constitution. A few of these promises have been met. There is free primary education. In 2007 the government issued a statement declaring that from 2008, secondary education would be heavily subsidised, with the government footing all tuition fees.

2007 Elections

The last general elections were held on December 27, 2007. In them, President Kibaki under the Party of National Unity ran for re-election against the main opposition party, the Orange Democratic Movement (ODM). The elections were seen to have been flawed with international observers saying that they were below international standards. After a split which would take a crucial 8% of the votes away from the ODM to the newly formed Orange Democratic Movement-Kenya (ODM-K)'s candidate, Kalonzo Musyoka, the race tightened between ODM candidate Raila Odinga and Kibaki. As the count came in to the Kenyan Election Commission, Odinga was shown to have a slight, and then substantial lead. However, as the Electoral Commission of Kenya (ECK) continued to count the votes, Kibaki closed the gap and then overtook his opponent by a substantial margin. This led to protests and open discrediting of the ECK for complicity and to Odinga declaring himself the "people's president" and calling for a recount. The protests escalated into violence and destruction of property. A group of eminent persons of Africa, led by former United Nations secretary-general Kofi Annan, brokered a peaceful solution to the political stalemate. This group enjoyed the backing of the UN, European Union, African Union and United States government, as well as those of various other notable countries across the world. More information is available in clashes in Kenya (2007–present). Annan requested mediation support for his team on the Panel Secretariat from the Swiss based conflict mediation organisation, the Centre for Humanitarian Dialogue.

2008

On 28 February 2008, Kibaki and Odinga signed an agreement on the formation of a coalition government in which Odinga would become Kenya's second prime Minister. Under the deal, the president would also appoint cabinet ministers from both PNU and ODM camps depending on each party's strength in Parliament. The agreement stipulated that the cabinet would also include a vice-president and two deputy Prime Ministers. After being debated and passed by Parliament, the coalition would hold until the end of the current Parliament or if either of the parties withdraws from the deal before then.

The new office of the PM will have power and authority to co-ordinate and supervise the functions of the Government and will be occupied by an elected MP who will also be the leader of the party or coalition with majority members in Parliament. The world watched Annan and his UN-backed panel and African Union chairman Jakaya Kikwete as they brought together the erstwhile rivals to the signing ceremony, beamed live on national TV from the steps of Nairobi's Harambee House. On February 29, 2008, representatives of PNU and ODM began working on the finer details of the power-sharing agreement. Kenyan lawmakers unanimously approved a power-sharing deal March 18, 2008, aimed at salvaging a country usually seen as one of the most stable and prosperous in Africa. The deal brought Kibaki's PNU and Odinga's ODM together and heralded the formation of the grand coalition, in which the two political parties would share power equally.

Grand Coalition

On April 13, 2008, President Kibaki named a grand coalition cabinet of 41 Ministers-including the prime minister and his two deputies. The cabinet, which also included 50 Assistant Ministers, was sworn in at the State House in Nairobi on Thursday, April 17, 2008 in the presence of Dr. Kofi Annan and other invited dignitaries. A constitutional change is being considered that will eliminate the position of Prime Minister and some powers will be transferred to the President. The changes will be debated in Parliament and a national referendum is necessary to cement the change.

Provinces, Districts and Divisions

Kenya comprises eight provinces each headed by a Provincial Commissioner (centrally appointed by the president). The provinces (*mkoa* singular *mikoa* plural in Swahili) are subdivided into districts (*wilaya*). There were 69 districts as of 1999 census. Districts are then subdivided into 497 divisions (*taarafa*). The divisions are then subdivided into 2,427 locations (*mtaa*) and then 6,612 sublocations (*mtaa mdogo*). The City of Nairobi enjoys the status of a full administrative province. The government supervises administration of districts and provinces. The provinces are:

1. Central
2. Coast
3. Eastern
4. Nairobi
5. North Eastern
6. Nyanza
7. Rift Valley
8. Western.

Local governance in Kenya is practised through local authorities. Many urban centres host city, municipal or town councils. Local authorities in rural areas are known as county councils. Local councillors are elected by civic elections, held alongside general elections. Constituencies are an electoral subdivision. There are 210 Constituencies in Kenya.

Population of Major Cities

City	*Population*
Nairobi	2,940,911
Mombasa	707,400
Nakuru	337,200
Kisumu	273,400
Eldoret	249,100
Nyeri	213,000
Machakos	179,500
Meru	140,900

Economy

Kenya's economy is market-based, with a few state-owned infrastructure enterprises, and maintains a liberalized external trade system. The country is generally perceived as Eastern and Central Africa's hub for Financial, Communication and Transportation services. As at May 2010, economic prospects are positive with 4-5% GDP growth expected, largely because of expansions in tourism, telecommunications, transport, construction and a recovery in agriculture. The World Bank predicts growth of 4% in 2010 and a potential of 4.9% growth in 2011. These improvements are supported by a large pool of English speaking professional workers. There is a high level of Computer literacy, especially among the youth.

The government is generally perceived as investment friendly and has enacted several regulatory reforms to simplify both foreign and local investment. An increasingly significant portion of Kenya's foreign inflows is from remittances by non-resident Kenyans who work in the US, Middle East, Europe and Asia. Compared to its neighbours, Kenya has a well developed social and physical infrastructure making it an attractive alternative location to South Africa, for major corporations seeking entry into the African continent.

In March 1996, the Presidents of Kenya, Tanzania, and Uganda re-established the East African Community (EAC). The EAC's objectives include harmonizing tariffs and customs regimes, free movement of people, and improving regional infrastructures. In March 2004, the three East African countries signed a Customs Union Agreement.

Nairobi continues to be the primary communication and financial hub of East Africa. It enjoys the region's best transportation linkages, communications infrastructure, and trained personnel. A wide range of foreign firms maintain regional branch or representative offices in the city.

Finance and Investment

Kenya is East and Central Africa's hub for Financial services. The Nairobi Stock Exchange (NSE) is ranked 4th in Africa in terms of Market capitalization. The Kenya banking system is supervised by the Central Bank of Kenya (CBK). As of late July 2004, the system consisted of 43 commercial banks (down from 48 in 2001), several non-bank financial institutions, including mortgage companies, four savings and loan associations, and several score foreign-exchange bureaus.

- The exchange rate of the Kenya shilling between 2003 and 2010 has averaged at about KSh74-78 to US$1.
- As at April 2010 Kenya's inflation rate is 3.65%.

Tourism

Kenya's services sector, which contributes about 63 percent of GDP, is dominated by tourism. The tourism sector has exhibited steady growth in most years since independence and by the late 1980s had become the country's principal source of foreign exchange. Tourists, the largest number from Germany and the United Kingdom, are attracted mainly to the coastal beaches and the game parks, notably, the expansive Tsavo National Park (20,808 square kilometres) in the southeast. Tourism has seen a substantial revival over the past several years and is the major contributor to the pick-up in the country's economic growth.

Tourism is now Kenya's largest foreign exchange earning sector, followed by flowers, tea, and coffee. In 2006 tourism generated US$803 million, up from US$699 million the previous year.

Agriculture

Agriculture is the second largest contributor to Kenya's gross domestic product (GDP), after the service sector. In 2005 agriculture, including forestry and fishing, accounted for about 24 percent of GDP, as well as for 18 percent of wage employment and 50 percent of revenue from exports. The principal cash crops are tea, horticultural produce, and coffee; horticultural produce and tea are the main growth sectors and the two most valuable of all of Kenya's exports. The production of major food staples such as corn is subject to sharp weather-related fluctuations. Production downturns periodically necessitate food aid—for example, in 2004 aid for 1.8 million people because of one of Kenya's intermittent droughts. Tea, coffee, sisal, pyrethrum, corn, and wheat are grown in the fertile highlands, one of the most successful agricultural production regions in

Africa. Livestock predominates in the semi-arid savanna to the north and east. Coconuts, pineapples, cashew nuts, cotton, sugarcane, sisal, and corn are grown in the lower-lying areas.

Industry and Manufacturing

Although Kenya is the most industrially developed country in East Africa, manufacturing still accounts for only 14 percent of gross domestic product (GDP). Industrial activity, concentrated around the three largest urban centres, Nairobi, Mombasa, and Kisumu, is dominated by food-processing industries such as grain milling, beer production, and sugarcane crushing, and the fabrication of consumer goods, e.g., vehicles from kits. There is a vibrant and fast growing cement production industry. Kenya also has an oil refinery that processes imported crude petroleum into petroleum products, mainly for the domestic market. In addition, a substantial and expanding informal sector engages in small-scale manufacturing of household goods, motor-vehicle parts, and farm implements.

Kenya's inclusion among the beneficiaries of the U.S. Government's African Growth and Opportunity Act (AGOA) has given a boost to manufacturing in recent years. Since AGOA took effect in 2000, Kenya's clothing sales to the United States increased from US$44 million to US$270 million (2006). Other initiatives to strengthen manufacturing have been the new government's favourable tax measures, including the removal of duty on capital equipment and other raw materials.

Energy

The largest share of Kenya's electricity supply comes from hydroelectric stations at dams along the upper Tana River, as well as the Turkwel Gorge Dam in the west. A petroleum-fired plant on the coast, geothermal facilities at Olkaria (near Nairobi), and electricity imported from Uganda make up the rest of the supply. Kenya's installed capacity stood at 1,142 megawatts a year between 2001 and 2003. The state-owned Kenya Electricity Generating Company (KenGen), established in 1997 under the name of Kenya Power Company, handles the generation of electricity, while the Kenya Power and Lighting Company (KPLC), which is slated for privatization, handles transmission and distribution. Shortfalls of electricity occur periodically, when drought reduces water flow. Kenya has yet to find hydrocarbon reserves on its territory, despite several decades of intermittent exploration.

Kenya currently imports all crude petroleum requirements. Petroleum accounts for 20 to 25 percent of the national import bill.

Other Industries

Other industries include forestry, fishing, and mining. Kenya has few discovered minerals and the mining industry is relatively minor. There is also an informal sector that, though generally not included in statistics, is increasingly becoming a very important contributor to the Kenya economy.

Vision 2030

In 2007, the Kenyan government unveiled Vision 2030, which is an economic blueprint which is believed to have the potential of putting the country in the same league as the Asian Economic Tigers.

Natural Gas Discovery

In the first quarter of 2010, China's CNOOC discovered unconfirmed but significant quantities of Natural Gas in North Eastern Kenya. An ongoing assessment will determine whether the gas quantities are commercially viable before a formal announcement is made. The gas discovery has raised hopes for that it is only a matter of time before Oil is discovered.

Donor Relations

Since independence, Kenya, a nonaligned country, has seen both substantial foreign investment and significant amounts of development aid, some from Russia, some from China and others from the West.

Kenya's development assistance has come from increasingly diverse sources in recent years with China taking a increasingly higher prominent role than the west. The share of funding provided by the United Kingdom has fallen significantly, while that of multilateral agencies, particularly the World Bank and the European Development Fund, has increased. The most active investors currently are the Chinese.

Criticism and Challenges

The economy's heavy dependence on rain-fed agriculture and the tourism sector leaves it vulnerable to cycles of boom and bust. The agricultural sector employs nearly 75 percent of the country's 38 million people. Half of the sector's output remains subsistence production. Poor governance and corruption also have had a negative impact on growth, making it expensive to do business in Kenya. According to Transparency International, Kenya ranks poorly in the corruption perception index though there have been significant improvements in recent times. (HIV/AIDS) continues to pose a long term risk to the economy. The Government has implemented awareness programmes to control its spread. Antiretroviral drugs are available in at government subsidized rates.

Kenya's economic policies have been subjected to persistent heavy criticism by western donors-especially when the policies do not ultimately benefit the West. Despite early disillusionment of western donors with the government, the economy has seen a broad-based expansion, led by strong performance in tourism and telecommunications, and acceptable post-drought results in agriculture, especially the vital tea sector. Kenya's economy grew by more than 7% in 2007 and its foreign debt was greatly reduced. Western donors are now adopting a less paternalistic attitude towards their relations with African nations. However there is still significant improvement to be done.

Demographics

Kenya's population has rapidly increased over the past several decades, and consequently it is relatively young. Some 73% of Kenyans are under 30. In 80 years, Kenya's population has grown from 2.9 million to 37 million.

Kenya is a country of great ethnic diversity. Most Kenyans are bilingual in English and Swahili, also a large percentage speak the mother tongue of their ethnic tribe.

Ethnic Groups

Kikuyu 22%, Luhya 14%, Luo 13%, Kalenjin 12%, Kamba 11%, Kisii 6%, Meru 6%, other African 15%, non-African (Asian, European, and Arab) 1%.

Largest Cities

Nairobi, Mombasa, Kisumu, Nakuru and Eldoret.

Health in Kenya

There are 3-5 births per woman on average. Life expectancy is estimated at between 47 and 55 years. There is a large number of HIV-positive people in Kenya. Maternal mortality is high, partly due to Female Genital Mutilation.

Education

Kenya's education system consists of early childhood education, primary, secondary and college. Early childhood education takes at least three years, primary eight years, secondary four and university four or six years depending on the course. Preschooling, which targets children from age three to five, is an integral component of the education system and is a key requirement for admission to Standard One (First Grade). At the end of primary education, pupils sit the Kenya Certificate of Primary Education (KCPE), which determines those who proceed to secondary school or vocational training. Primary school age is 6/7-13/14 years. For those who proceed to secondary level, there is a national examination at the end of Form Four – the Kenya Certificate of Secondary Education (KCSE), which determines those proceeding to the universities, other professional training or employment. The Joint Admission Board (JAB) is responsible for selecting students joining the public universities. Other than the public schools, there are many private schools in the country, mainly in urban areas. Similarly, there are a number of international schools catering for various overseas educational systems.

Historical Background

Independent Kenya's first system of education was introduced by British colonists. After Kenya's independence on December 12, 1963, an authority named Ominde Commission was formed to introduce changes that would reflect the nation's sovereignty. The commission focused on identity and unity, which were critical issues at the time. Changes in the subject content of history and geography were made to reflect national cohesion. Between 1964 and 1985, the 7-4-2-3-system was adopted – seven years of primary, four years of lower secondary, two years of upper secondary, and three years of university. All schools had a common curriculum.

In 1981, the Presidential Working Party on the Second University was commissioned to look at both the possibilities of setting up a second university in Kenya as well as the reforming of the entire education system. The committee recommended that the 7-4-2-3 system be changed to an 8-4-4 system (eight years in primary, four years in secondary, and four years in university education). The table under Present-day education in Kenya below shows the structure of the 8-4-4 system. Although the 7-4-2-3 system theoretically ended with the introduction of the new 8-4-4 system in 1985, the last batch of students from the former system graduated from Kenyan Universities in 1992.

Present-day Education in Kenya

The current 8-4-4 system was launched in January 1985. It put more emphasis on vocational subjects on the assumption that the new structure would enable school dropouts at all levels either to be self-employed or to secure employment in the informal sector.

In January 2003, the Government of Kenya announced the introduction of free primary education. As a result, primary school enrolment increased by about 70%. However, secondary and tertiary education enrollment has not increased proportionally because payment is still required for attendance.

In class eight of primary school the Kenya Certificate of Primary Examination (K.C.P.E.) is written. The result of this examination is needed for placement at secondary school. In form four of secondary schools the Kenya Certificate of Secondary Examination (K.C.S.E.) is written. Students sit examinations in eight subjects.

KCSE Grading System

Grade	A	A-	B+	B	B-	C+	C	C-	D+	D	D-	E
Points	12	11	10	9	8	7	6	5	4	3	2	1

The average grade is based on performance in the eight subjects. Where a candidate sits for more than eight subjects, the average grade is based on the best eight subjects. University matriculation is based on the best eight and performance in particular subjects relevant to degree courses. Example below:

Subject	*Group*	*Grade*	*Points*
English	1	B+	10
Kishwahili	1	A-	11
Mathematics	1	A	12
History & Government	3	B	9
Geography	3	A-	11
Physics	2	B+	10
Chemistry	2	B-	8
Biology	2	A-	11

The total number of points is 81.

The average grade is 81 divided by 8, which equals 10.1 (approximately 10.0 points) which is Grade B+ according to the grading system. This student qualifies to join one of the Public Universities for his good score. Training institutions and faculties and departments determine their own minimum entry requirements.

Students who manage a grade of C+ qualify to do a degree course at the University. Owing to competition, and fewer places at the University, those with B and in a few cases B- and above are taken for degree courses at the Public Universities and benefit by paying government-subsidised fees. The rest join private universities or middle-level colleges.

Interestingly, the number of students admitted to public universities through J.A.B depends on the total number of beds available in all the public universities.

Nonetheless, those who miss out but attained the minimum university entry mark of C+ or C with a relevant diploma certificate are admitted through the parallel degree programmes (module II) if they can afford the full fees for the course.

This has been the subject of much discussion with people questioning the rationale and morality of locking out qualified students from public institutions yet still admitting those who come from financially able families.

Criticism

The Kenyan 8-4-4 system of education has weathered many storms in the 24 years of its existence. Immediately after the first batch of students of this system graduated in 1989, most of the country was up in arms, criticizing the government of "haphazard adoption of the system".

Critics claimed that the system was producing "below par" graduates who could not effectively compete with their counterparts from other parts of the world. Others argued that the system had produced graduates who were either too young or ill-prepared for the job market. The government turned a deaf ear, probably because the prospect of overhauling the new system was unwelcome for the cost that the exercise would involve.

Over the years however graduates have proved critics wrong by excelling in universities locally and abroad. Indeed, the braindrain being experienced where health workers and scholars emigrate to Western countries is proof of this. Many 8-4-4 graduates have also excelled in universities outside the country.

The emphasis on vocational training has waned and recent changes to the curriculum have now laid more emphasis on information technology, sciences, mathematics and languages. In any case, the academic workload and emphasis on passing written examinations has left little room for carpentry, masonry, cooking and other vocational training.

Culture

Kenya is a diverse country, with many different cultures represented. Notable cultures include the Swahili on the coast, pastoralist communities in the north, and several different communities in the central and western regions.

Today, the Maasai culture is well known, given its heavy exposure from tourism, however, Maasai make up a relatively minor percentage of the Kenyan population. The Maasai are known for their elaborate upper body adornment and jewelry.

Kenya has an extensive music, television and theatre scene.

Literature

Ngugi wa Thiong'o is one of the best known writers of Kenya. His book, *Weep Not, Child* is an illustration of life in Kenya during the British occupation. This is a story about the effects of the Mau Mau on the lives of black Kenyans. Its combination of themes—colonialism, education, and love—helped to make it one of the best-known novels in Africa.

M.G. Vassanji's 2003 novel *The In-Between World of Vikram Lall* won the Giller Prize in 2003. It is the fictional memoir of a Kenyan of Indian heritage and his family as they adjust to the changing political climates in colonial and post-colonial Kenya.

Since 2003, the literary journal *Kwani?* has been publishing Kenyan contemporary literature.

Sports

Kenya is active in several sports, among them cricket, rallying, football (soccer), rugby union and boxing. But the country is known chiefly for its dominance in Middle-distance and long-distance athletics. Kenya has consistently produced Olympic and Commonwealth Games champions in various distance events, especially in 800 m, 1,500 m, 3,000 m steeplechase, 5,000 m, 10,000 m and the marathons. Kenyan athletes (particularly Kalenjin) continue to dominate the world of distance running, although competition from Morocco and Ethiopia has reduced this supremacy. Kenya's best-known athletes included the four-time women's Boston Marathon winner and two-time world champion Catherine Ndereba, former Marathon world record-holder Paul Tergat, and John Ngugi.

Kenya won several medals during the Beijing Olympics, 5 gold, 5 silver and 4 bronze, making it Africa's most successful Nation in the 2008 Olympics. New athletes gained attention, such as Pamela Jelimo, the women's 800m gold medalist who went ahead to win the IAAF Golden League jackpot, and Samuel Wanjiru who won the men's marathon.

Retired Olympic and Commonwealth Games champion Kipchoge Keino helped usher in Kenya's ongoing distance dynasty 1970s and was followed by Commonwealth Champion Henry Rono's spectacular string of world record performances.

Lately, there has been controversy in Kenyan athletics circles, with the defection of a number of Kenyan athletes to represent other countries, chiefly Bahrain and Qatar. The Kenyan Ministry of Sports has tried to stop the defections, but they have continued anyway, with Bernard Lagat the latest, choosing to represent the United States. Most of these defections occur because of economic or financial factors however some elite Kenyan runners who cannot qualify for their country's strong national team also find it easier to qualify by running for other countries.

Kenya has also been a dominant force in ladies' volleyball within Africa, with both the clubs and the national team winning various continental championships in the past decade. The women's team has also competed at the Olympics and World Championships but without any notable success.

Cricket is another popular and the most successful team sport. Kenya has competed in the Cricket World Cup since 1996. They upset some of the World's best teams and reached semifinals of the 2003 tournament. They also won the inaugural World Cricket League Division 1 hosted in Nairobi and participated in the World T20. Their current captain is Steve Tikolo.

Kenya is represented by Lucas Onyango as a professional rugby league player, plying his trade with Oldham Roughyeds. Besides the former European Super League team, he has also played for Widnes Vikings and rugby union with Sale Sharks.

Kenya is making a name for itself in rugby union. It is popular in Kenya especially with the annual Safari Sevens tournament. Kenya sevens team ranked 9th in IRB Sevens World Series for the 2006 season.

Kenya was a regional power in soccer but its dominance has been eroded by wrangles within the Kenya Football Federation. This has led to a suspension by FIFA which was lifted in March, 2007.

In the motor rallying arena, Kenya is home to the world famous Safari Rally, commonly acknowledged as one of the toughest rallies in the world, and a part of the World Rally Championship for many years until its exclusion after the 2002 event owing to financial difficulties.

Some of the best rally drivers in the world have taken part in and won the rally, such as Bjorn Waldegard, Hannu Mikkola, Tommi Makinen, Shekhar Mehta, Carlos Sainz and Colin McRae. Though the rally still runs annually as part of the Africa rally championship, the organisers are hoping to be allowed to rejoin the World Rally championship in the next couple of years.

4

The Islamic Republic of Mauritania

Mauritania, officially the Islamic Republic of Mauritania, is a country in North Africa. It is bordered by the Atlantic Ocean in the west, by Western Sahara in the north, by Algeria in the northeast, by Mali in the east and southeast, and by Senegal in the southwest. It is named after the Roman province of Mauretania, even though the modern state covers a territory far to the southwest of the old province. The capital and largest city is Nouakchott, located on the Atlantic coast.

The civilian government of Mauritania was overthrown on 6 August 2008, in a military *coup d'etat* led by General Mohammed Ould Abdel Aziz. On 16 April 2009, General Aziz resigned from the military to run for president in the 19 July elections, which he won. In Mauritania about 20% of the population live on less than US $1.25 per day.

History

The Bafours were primarily agriculturalist, and among the first Saharan people to abandon their historically nomadic lifestyle. With the gradual desiccation of the Sahara, they headed south.

Following them came a migration of not only Central Saharans into West Africa, but in 1076, Moorish Islamic warrior monks (Almoravid or Al Murabitun) attacked and conquered the ancient Ghana Empire. Over the next 500 years, Arabs overcame fierce resistance from the local population (Berber and non-Berber alike) and came to dominate Mauritania. The Mauritanian Thirty-Year War (1644–74) was the unsuccessful final effort to repel the Yemeni Maqil Arab invaders led by the Beni Hassan tribe.

The descendants of the Beni Hassan warriors became the upper stratum of Moorish society. Berbers retained influence by producing the majority of the region's Marabouts—those who preserve and teach Islamic tradition. Many of the Berber tribes claimed Yemeni (and sometimes other Arab) origin: there is little evidence to suggest this, though some studies do make a connection between the two. Hassaniya, a Berber-influenced Arabic dialect that derives its name from the Beni Hassan, became the dominant language among the largely nomadic population.

Imperial France gradually absorbed the territories of present-day Mauritania from the Senegal river area and upwards, starting in the late 1800s. In 1901, Xavier Coppolani took charge of the imperial mission. Through a combination of strategic alliances with Zawiya tribes, and military pressure on the Hassane warrior nomads, he managed to extend French rule over the Mauritanian emirates: Trarza, Brakna and Tagant quickly submitted to treaties with the colonial power (1903–04), but the northern emirate of Adrar held out longer, aided by the anticolonial rebellion (or jihad) of shaykh Maa al-Aynayn. It was finally defeated militarily in 1912, and incorporated into the territory of Mauritania, which had been drawn up in 1904. Mauritania would subsequently form part of French West Africa, from 1920.

French rule brought legal prohibitions against slavery, and an end to interclan warfare. During the colonial period, the population remained nomadic, but many sedentar peoples, whose ancestors had been expelled centuries earlier, began to trickle back into Mauritania. As the country gained independence in 1960, the capital city Nouakchott was founded at the site of a small colonial village, the Ksar, while 90% of the population was still nomadic.

The great Sahel droughts of the early 1970s caused massive problems in Mauritania. With independence, larger numbers of indigenous Sub-Saharan African peoples (Haalpulaar,

Soninke, and Wolof) entered Mauritania, moving into the area north of the Senegal River. Educated in French language and customs, many of these recent arrivals became clerks, soldiers, and administrators in the new state. This occurred as France militarily suppressed the most intransigent Hassane tribes of the Moorish north, shifting old balances of power, and creating new cause for conflict between the southern populations and Moors. Between these groups stood the Haratin, a very large population of Arabized slaves of black African origins, who lived within Moorish society, integrated into a low-caste social position. Modern day slavery is still a common practice in this country. According to some estimates, up to 600,000 Mauritanians, or 20% of the population, are still enslaved. This social discrimination concerns mainly the Haratin ("black Moors) in the northern part of the country, where tribal elites among "white Moors" (Beidane) hold sway, but low-caste groups within the black African communities of the south are also affected by similar practices.

Moors reacted to the change, and to Arab nationalist calls from abroad, by increasing pressure to Arabize many aspects of Mauritanian life, such as law and language. A schism developed between those Moors who consider Mauritania to be an Arab country and those who seek a dominant role for the non-Moorish peoples, with various models for containing the country's cultural diversity suggested, but none implemented successfully.

This ethnic discord was evident during intercommunal violence that broke out in April 1989 (the "1989 Events" and "Mauritania-Senegal Border War"), but has since subsided. Some 70,000 black African Mauritanians were expelled from Mauritania in the late 1980s. The ethnic tension and the sensitive issue of slavery – past and, in some areas, present – is still a powerful theme in the country's political debate. A significant number from all groups, however, seek a more diverse, pluralistic society.

The government bureaucracy is composed of traditional ministries, special agencies, and parastatal companies. The Ministry of Interior spearheads a system of regional governors and prefects modelled on the French system of local administration. Under this system, Mauritania is divided into thirteen regions (*wilaya*), including the capital district, Nouakchott. Control is tightly concentrated in the executive branch of the central government, but a series of national and municipal elections since 1992 have produced limited decentralization.

Mauritania, along with Morocco, annexed the territory of Western Sahara in 1976, with Mauritania taking the lower one-third at the request of former imperial power Spain. After several military losses to the Polisario – heavily armed and supported by Algeria, the local hegemon and rival to Morocco – Mauritania retreated in 1979, and its claims were taken over by Morocco. Due to economic weakness, Mauritania has been a negligible player in the territorial dispute, with its official position being that it wishes for an expedient solution that is mutually agreeable to all parties. While most of Western Sahara has been occupied by Morocco, the UN still considers the Western Sahara a territory that needs to express its wishes with respect to statehood: a referendum is still supposed to be held sometimes in the future, under UN auspices, to determine whether or not the indigenous Sahrawis wish to independent as the Sahrawi Arab Democratic Republic, or to be part of Morocco. The Moroccan Government has thus far blocked such a referendum from taking place.

The Ould Daddah Era (1960-78)

After independence, President Moktar Ould Daddah, originally installed by the French, formalized Mauritania into a one-party state in 1964 with a new constitution, which set up an authoritarian presidential regime. Daddah's own Parti du Peuple Mauritanien (PPM) became the ruling organization in a single-party system. The President justified this decision on the grounds that he considered Mauritania unready for western-style multi-party democracy.

Under this one-party constitution, Daddah was re-elected in uncontested elections in 1966, 1971 and 1976. He was ousted in a bloodless coup on 10 July 1978, after bringing the country to near-collapse through a disastrous war to annex the southern part of Western Sahara, in an attempt to create a "Greater Mauritania".

CMRN and CMSN Military Governments (1978-84)

Col. Mustafa Ould Salek's CMRN *junta* proved incapable of either establishing a strong base of power or extracting the country from its destabilizing conflict with the Sahrawi resistance movement, the Polisario Front. It quickly fell to be replaced by another military government, the CMSN. The energetic Col. Mohammed Khouna Ould Haidallah soon emerged as its main strongman, and by giving up all claims to Western Sahara, he found peace with the Polisario and improved relations with its main backer, Algeria – but relations with the other party to the conflict, Morocco, and its European ally France, deteriorated. Instability continued, and Haidallah's ambitious reform attempts foundered. Not only was his regime plagued by attempted coups and intrigue within the military establishment, but it also became increasingly contested because of his harsh and uncompromising line against opponents and political and military dissidents, of whom many were jailed and some were executed.

In 1984 he was deposed by Col. Maaouya Ould Sid Ahmed Taya, who relaxed the political climate somewhat, without relinquishing military control. Ould Taya moderated Mauritania's previous pro-Algerian stance, and reconnected with Morocco during the late 1980s. Relations with Morocco deepened during the late 1990s and early 2000s, as part of Mauritania's drive to attract support from Western states and Western-aligned Arab states. However, Mauritania has not rescinded its recognition of Polisario's Western Saharan exile government, remaining on good terms with Algeria. Its position on the Western Sahara conflict is, since the 1980s, one of strict neutrality.

Ould Taya's Rule (1984-2005)

The *Parti Republican Democratique et Social* (PRDS), formerly led by President Maaouya Ould Sid Ahmed Taya, dominated Mauritanian politics following the country's first multi-party elections in April 1992 following the approval by referendum of the current constitution in July 1991. President Taya, who won elections in 1992 and 1997, first became chief of state through a 12 December 1984 bloodless coup which made him chairman of the committee of military officers that governed Mauritania from July 1978 to April 1992. Political parties, illegal during the military period, were legalized again in 1991. By April 1992, as civilian rule returned, 16 major political parties had been recognized; 12 major political parties were active in 2004. Most opposition parties boycotted the first legislative election in 1992, and for nearly a decade the parliament was dominated by the PRDS. The opposition participated in municipal elections in January-February 1994 and subsequent Senate elections, most recently in April 2004, gained representation at the local level as well as three seats in the Senate.

A group of current and former Army officers launched a bloody but unsuccessful coup attempt on 8 June 2003. The leaders of the attempted coup were never caught.

Mauritania's presidential election, its third since adopting the democratic process in 1992, took place on 7 November 2003. Six candidates, including Mauritania's first female and first Haratine (former slave family) candidates, represented a wide variety of political goals and backgrounds. Incumbent President Maaouya Ould Sid Ahmed Taya won re-election with 67.02% of the popular vote, according to the official figures, with Mohammed Khouna Ould Haidalla finishing second.

During the late 1980s, Ould Taya had established a close co-operation with Iraq, and pursued a strongly Arab nationalist line. At the same time, bloody clashes erupted with Senegal in 1989, during which both countries expelled ethnic minorities to the other country. Mauritania grew increasingly isolated internationally, and tensions with Western countries grew dramatically after it took a pro-Iraqi position during the 1991 Gulf War. During the mid-to late 1990s, Mauritania shifted its foreign policy to one of increased co-operation with the US and Europe, and was rewarded with diplomatic relaxation and aid projects.

In 1999, Mauritanian Foreign Minister Ahmed Sid Ahmed and his Israeli counterpart David Levy signed an agreement in Washington DC, USA, on 28 October, establishing full diplomatic relations with Mauritania. The signing ceremony was held at the U.S. State Department in the presence of U.S. Secretary of State Madeleine Albright. Mauritania thereby joined Egypt, Palestine, and Jordan as the only members of the Arab League to officially recognize Israel. Ould Taya also started co-operating with the United States in antiterrorism activities, which was criticized by human rights NGOs, who talked of an exaggeration and instrumentation of alleged terrorist activities for geopolitical aims.

August 2005 military coup

On 3 August 2005, a military coup led by Colonel Ely Ould Mohammed Vall ended Maaouya Ould Sid Ahmed Taya's twenty-one years of rule.

On 3 August, the Mauritanian military, including members of the presidential guard, seized control of key points in the capital of Nouakchott. They took advantage of President Taya's attendance at the funeral of Saudi King Fahd to organize the coup, which took place without loss of life. The officers, calling themselves the Military Council for Justice and Democracy, released the following statement:

"The national armed forces and security forces have unanimously decided to put a definite end to the oppressive activities of the defunct authority, which our people have suffered from during the past years."

The Military Council later issued another statement naming Colonel Vall as president and director of the national police force, the *Surete National*. Sixteen other officers were listed as members. Colonel Vall was once regarded as a firm ally of the now-ousted president, even aiding him in the original coup that brought him to power, and later serving as his security chief.

Applauded by the Mauritanian people, but cautiously watched by the international community, the coup has since been generally accepted, while the military *junta* has organized elections within the promised two-year timeline. In a referendum on 26 June 2006, Mauritanians overwhelmingly (97%) approved a new constitution which limited the duration of a president's stay in office. The leader of the *junta*, Col. Vall, promised to abide by the referendum and relinquish power peacefully. Mauritania's establishment of relations with the State of Israel – it was one of only three Arab states to recognize Israel – was maintained by the new regime, despite widespread criticism from the opposition, who viewed it as a legacy of the Taya regime's attempts to curry favour with the West.

Parliamentary and municipal elections in Mauritania took place on 19 November and 3 December 2006.

2007 Presidential Election

The first fully democratic Presidential election since 1960 occurred on 11 March 2007. The election effected the final transfer from military to civilian rule following the military coup in 2005. This was the first time that the president had been selected in a multi-candidate election in the country's post-independence history.

The election was won in a second round of voting by Sidi Ould Cheikh Abdallahi, with Ahmed Ould Daddah a close second.

2008 Military Coup

The head of the Presidential Guards took over the president's palace and units of the army surrounded a key state building in the capital Nouakchott on 6 August 2008, a day after 48 lawmakers from the ruling party resigned. The army surrounded the state television building after the president sacked (fired) two senior officers, including the head of the presidential guards. The president, the prime minister and the minister of internal affairs were arrested.

The coup was organized by General Mohammed Ould Abdel Aziz, former chief of staff of the Mauritanian army and head of the Presidential Guard, whom the president had just dismissed. Mauritania's presidential spokesman, Abdoulaye Mamadouba, said President Sidi Ould Cheikh Abdallahi, Prime Minister Yahya Ould Ahmed Waghf and the interior minister, were arrested by renegade Senior Mauritanian army officers, unknown troops and a group of generals, and were held under house arrest at the presidential palace in Nouakchott. In the apparently successful and bloodless coup d'etat, Abdallahi's daughter, Amal Mint Cheikh Abdallahi, said: "The security agents of the BASEP (Presidential Security Battalion) came to our home and took away my father." The coup plotters, all dismissed in a presidential decree shortly beforehand, included General Muhammad Ould 'Abd Al-'Aziz, General Muhammad

Ould Al-Ghazwani, General Philippe Swikri, and Brigadier General (Aqid) Ahmad Ould Bakri.

After the Coup

A Mauritanian lawmaker, Mohammed Al Mukhtar, announced that "many of the country's people were supporting the takeover attempt and the government was "an authoritarian regime" and that the president had "marginalized the majority in parliament." The coup was also backed by Abdellahi's rival in the 2007 election, Ahmed Ould Daddah. However, Ould `Abd Al-`Aziz's regime was isolated internationally and punished by diplomatic sanctions and the cancellation of some aid projects. It found few supporters, among them Morocco, Libya and Iran, while Algeria, the United States, France and other European countries criticized the coup, and continued to refer to Abdellahi as the legitimate president of Mauritania. A group of parties also coalesced around Abdellahi to continue to protest the coup, causing the junta to ban demonstration and crack down on opposition activists. International and internal pressure eventually forced the release of Abdellahi, who was instead placed in house arrest in his home village. The new government broke off relations with Israel, which it had recognized in 1999, during the Gaza invasion of late 2008/early 2009, which helped grant it some recognition in the Arab world and promises of Iranian and Libyan support. Even so, the `Abd Al-`Aziz government appeared isolated and weak during the first half of 2009. In March 2010 Mauritania's female foreign minister Naha Mint Hamdi Ould Mouknass announced that Mauritania had cut ties with Israel in a "complete and definitive way."

'Abd Al-'Aziz had since the coup insisted on organizing new presidential elections to replace Abdellahi, but was forced to reschedule them due to internal and international opposition. However, during the spring of 2009, the junta negotiated an understanding with some opposition figures as well as international parties, which dramatically changed the situation. Abdellahi formally resigned, under protest, as it became clear that some opposition forces had defected from him and most international players, notably including France and Algeria, now lined up behind 'Abd Al-'Aziz. The United States continued to criticize the coup, but did not actively oppose the elections.

Abdellahi's resignation paved the way for the election of military strongman Muhammad Ould 'Abd Al-'Aziz as civilian president, on 18 July, by a 52% majority. Many of Abdellahi's former supporters criticized this as a political ploy and refused to recognize the results.

They argued that the election had been falsified due to junta control, and complained that the international community had let down the opposition. Despite marginal complaints, the elections were almost unanimously accepted by Western, Arab and African countries, which lifted sanctions and resumed cooperation with Mauritania.

By late summer, 'Abd Al-'Aziz appeared to have secured his position and to have garnered widespread international and internal support, although several influential parties and political personalities, notably Senate chairman Messaoud Ould Boulkheir, continued to refuse the new order and call for 'Abd Al-'Aziz's resignation.

Regions and Departments

Mauritania is divided into 12 regions (*regions*) called wilaya and one capital district in Nouakchott, which in turn are subdivided into 44 departments (*moughataa*). The regions and capital district (in alphabetical order) and their capitals are:

Region	*Capital*
Adrar	Atar
Assaba	Kifa
Brakna	Aleg
Dakhlet Nouadhibou	Nouadhibou
Gorgol	Kaedi
Guidimaka	Selibaby
Hodh Ech Chargui	Nema
Hodh El Gharbi	Ayoun el Atrous
Inchiri	Akjoujt
Nouakchott (capital district)	Nouakchott
Tagant	Tidjikdja
Tiris Zemmour	F'derik
Trarza	Rosso

Geography

At 397,929 square miles (1,030,631 km^2), Mauritania is the world's 29th-largest country (after Bolivia). It is comparable in size to Egypt.

Mauritania is generally flat, its 1,030,631 square kilometres (397,929 sq mi) forming vast, arid plains broken by occasional ridges and clifflike outcroppings. A series of scarps face southwest, longitudinally bisecting these plains in the centre of the country. The scarps also separate a series of sandstone plateaus, the highest of which is the Adrar Plateau, reaching an elevation of 500 meters (1,640 ft). Spring-fed oases lie at the foot of some of the scarps. Isolated peaks, often rich in minerals, rise above the plateaus; the smaller peaks are called guelbs and the larger ones kedias. The concentric Guelb er Richat (also known as the Richat Structure) is a prominent feature of the north-central region. Kediet ej Jill, near the city of Zouirat, has an elevation of 1,000 meters (3,281 ft) and is the highest peak.

Approximately three quarters of Mauritania is desert or semidesert. As a result of extended, severe drought, the desert has been expanding since the mid-1960s. To the west, between the ocean and the plateaus, are alternating areas of clayey plains (regs) and sand dunes (ergs), some of which shift from place to place, gradually moved by high winds. The dunes generally increase in size and mobility toward the north.

Economy

Economy-overview: A majority of the population of Mauritania still depends on agriculture and livestock for a livelihood, even though most of the nomads and many subsistence farmers were forced into the cities by recurrent droughts in the 1970s and 1980s. Mauritania has extensive deposits of iron ore, which account for almost 50% of total exports. The decline in world demand for this ore, however, has led to cutbacks in production. With the current rise in metal prices, gold and copper mining companies are opening mines in the interior. The nation's coastal waters are among the richest fishing areas in the world, but over exploitation by foreigners threatens this key source of revenue. The country's first deep water port opened near Nouakchott in 1986. In recent years, drought and economic mismanagement have resulted in a buildup of foreign debt. In March 1999, the government signed an agreement with a joint World Bank-IMF mission on a $54 million enhanced structural adjustment facility (ESAF). The economic objectives have been set for 1999-2002. Privatization remains one of the key issues.

Macro-economic Trend

This is a chart of trend of gross domestic product of Mauritania at market prices estimated by the International Monetary Fund with figures in millions of Mauritanian Ougulyas.

Current GDP per capita of Mauritania grew 82% in the Sixties reaching a peak growth of 166% in the Seventies. But this proved unsustainable and growth consequently scaled back to 14% in the Eighties. Finally, it shrank by 29% in the Nineties.

Average wages in 2007 hover around $4-5 per day.

Diversification

In 2007, mining industries accounted for well over 35 per cent of the Mauritanian economy, with the fish industry so much as 54% (with big changes between these industries in the power relationship). Diversification of the economy into non-mining industries remains a long-term issue. Mauritania is a net importer of food, reportedly importing 70% of its domestic food needs.

Dispute with Wood Side Petroleum

In February 2006, the Mauritanian government denounced amendments to an oil contract made by former leader Maaouiya Ould Taya with Wood side Petroleum, an Australian company. In 2004, Wood side had agreed to invest $US 600 million in developing Mauritania's Chinguetti offshore oil project. The controversial amendments, which Mauritanian authorities declared had been signed "outside the legal framework of normal practice, to the great detriment of our country", could cost Mauritania up to $200 million a year, according to *BBC News*. Signed by Wood side two weeks after the February 1, 2005 legislation authorizing the four amendments, they provided for a lower state quota in the *profit-oil*, and reduced taxes by 15 percent in certain zones. They also eased environmental constraints, and extended the length and scope of the exploitation and exploration monopoly, among other measures.

The disputed amendments were signed by former oil minister Zeidane Ould Hmeida in February 2004 and March 2005. Hmeida was arrested in January 2006 on charges of "serious crimes against the country's essential economic interests".

Nouakchott's authorities declared that the government would likely seek international arbitration, which Wood side (which operated for Hardman, BG Group, Premier, ROC Oil, Fusion, Petronas, Dana Petroleum, Energy Africa and the Hydrocarbons Mauritanian Society) also contemplated.

Discovered in 2001, Chinguetti has proven reserves of about 120 million barrels of oil. At the end of December 2005, authorities estimated that in 2006, the oil profits would be 47 billion *ouguiyas* (about US$180 million) and represent a quarter of the state budget, according to *RFI*.

The Australian Federal Police are currently investigating Wood side for allegations of bribery and corruption in Mauritania (according to the Sydney Morning Herald).

Some U.S. Oil companies are alleged to be playing a part in Mauritania's oil related corruption.

Mauritania has one of the lowest GDP rates in Africa, despite being rich in natural resources. However, a majority of the population still depends on agriculture and livestock for a livelihood, even though most of the nomads and many subsistence farmers were forced into the cities by recurrent droughts in the 1970s and 1980s. Mauritania has extensive deposits of iron ore, which account for almost 50% of total exports. With the current rises in metal prices, gold and copper mining companies are opening mines in the interior. The nation's coastal waters are among the richest fishing areas in the world, but over exploitation by foreigners threatens this key source of revenue. The country's first deepwater port opened near Nouakchott in 1986. In recent years, drought and economic mismanagement have resulted in a buildup of foreign debt. In March 1999, the government signed an agreement with a joint

World Bank-IMF mission on a $54 million enhanced structural adjustment facility (ESAF). The economic objectives have been set for 1999-2002. Privatization remains one of the key issues. Mauritania is unlikely to meet ESAF's annual GDP growth objectives of 4%-5%.

Oil was discovered in Mauritania in 2001 in the offshore Chinguetti deposit. Although potentially significant for the Mauritanian economy, it remains to be seen how much it will help the country. Mauritania has been described as a "desperately poor desert nation, which straddles the Arab and African worlds and is Africa's newest, if small-scale, oil producer." There may be additional oil reserves inland in the Taoudeni basin, although the harsh environment will make extraction expensive.

The Government's main problem is currently privatizing the economy.

Demographics

Population : 3,364,940 (July 2008 estimated)

Ethnic groups : 40% mixed Moor/Black; 30% Moor; 29% Black; 1% French

Religion : near 100% Muslim, most of whom are Sunnis.

Languages : Hassaniya dialect of Arabic (official and national);

Other languages spoken include: Pulaar, Soninke, Imraguen language, Wolof and French (widely used in media and among educated classes, see African French).

Health

Life expectancy at birth was 53.91 years (2008 estimate). Per capita expenditure on health was 43 US$ (PPP) in 2004. Public expenditure was 2% of the GDP in 2004 and private 0.9 % of the GDP in 2004. In the early 21st century there were 11 physicians per 100,000 people. Infant mortality was 7,8 % of the live births.

Culture

The name of the country is derived from the Latin Mauretania, meaning the land of the Mauri.

The French occupied the country in 1860 in close cooperation with Maur religious leaders. Mauritania became a nation after the destruction of the kingdoms of Fouta Toro and Walo Walo and the Arab-Berber emirates of Trarza, Brakna, Taganet, and Adrar. As a result, the country has two main ethnic groups: black Africans and Arab-Berbers. The black African group includes the Fulani, Soninke, and Bambara. The Maurs include the Arab-Berbers (Beydan) and the black Maurs known as Haratin. The Haratins are black Africans who were enslaved by white Maurs. White and black Maurs consider themselves Arab, whereas non-Arab blacks see themselves as African. The most important common denomination, if not the only one, is Sunni Islam.

- Music of Mauritania
- Islam in Mauritania
- Status of religious freedom in Mauritania
- Mauritania and Madagascar are the only two countries in the world not to use decimal-based currency. The basic unit of currency, the ouguiya, comprises five khoums
- Filming for several documentaries and films has taken place in Mauritania, including *Fort Saganne* (1984), *The Fifth Element* (1997), *The Books Under the Sand* (1997), *Life without Death* (1997), *Winged Migration* (2001), and *Heremakono* (2002.

Education

Since 1999, all teaching in the first year of primary school is in Arabic, French is however introduced in the second year and all scientific courses are taught in French for everyone. The use of English and the Weldiya dialect is also on the increase. The country has the University of Nouakchott and other institutions of higher education. Public expenditure on education was at 10.1 % of 2000-2007 government expenditure.

5

The Kingdom of Morocco

Morocco, officially the Kingdom of Morocco, is a country located in North Africa. It has a population of nearly 32 million and an area of 710.850 km², including the disputed Western Sahara which is mainly under Moroccan administration. Morocco has a coast on the Atlantic Ocean that reaches past the Strait of Gibraltar into the Mediterranean Sea. It is bordered by Spain to the north (a water border through the Strait and land borders with three small Spanish exclaves, Ceuta, Melilla, and Peñón de Vales de la Gomera), Algeria to the east, and Mauritania to the south.

Morocco is a de jure constitutional monarchy with an elected parliament. The King of Morocco holds vast executive powers, including dissolving parliament. Executive power is exercised by the government. Legislative power is vested in both the government and the two chambers of parliament, the Assembly of Representatives and the Assembly of Councillors. Parliamentary elections were held in Morocco on 7 September 2007, and were considered by observers as mostly free and fair; although voter turnout was estimated to be 37%, the lowest in Moroccan political history. The political capital is Rabat, and the largest city is Casablanca; other large cities includes Fes, Sale, Agadir, Marrakesh, Tangier, Meknes, Oujda and Tetouan.

The history of Morocco spans over 12 centuries, and the country was first created by the Idrisid dynasty, whose origins are from Saudi Arabia, representing the first Islamic state in Africa autonomous from the Arab Empire. Under the Almoravid dynasty and the Almohad dynasty, Morocco dominated the

Maghreb and Muslim Spain. The Reconquista ended Almohad rule in Iberia and many Muslims and Jews migrated to Morocco. Under the Saadi Dynasty, Morocco would consolidate power and fight off Portuguese and Ottoman invaders, as in the battle of Ksar el Kebir. The reign of Ahmad al-Mansur brought new wealth and prestige to the Sultanate, and an invasion of the Songhay Empire was initiated. However, managing the territories across the Sahara proved too difficult. After the death of al-Mansur the country was divided among his sons. In 1666 the sultanate was reunited by the Alaouite dynasty, who have since been the ruling house in Morocco. The organization of the state developed with Ismail Ibn Sharif. With his Black Guard he drove the English from Tangier (1684) and the Spanish from Larache (1689). The Alaouite dynasty distinguished itself in the 20th century by maintaining Moroccan independence while other states in the region succumbed to European interests. In 1912, after the First Moroccan Crisis and the Agadir Crisis, the Treaty of Fez was signed, effectively dividing Morocco into a French and Spanish protectorate. In 1956, after 44 years of occupation Morocco regained indepedence from France as the Kingdom of Morocco.

Morocco has a rich culture and civilization, which remained mainly indigenous throughout times and the Moroccan cuisine has long been considered as one of the most diversified cuisines in the world. The population is almost entirely Arab-Berber. Although Arabic is the majority language, modern studies show that the Arabization process in Morocco was mostly cultural. The Moroccans or Moroccan Arabs, are a largely homogenous group speaking Moroccan Arabic, although regional variation does occur. The Berber people in Morocco can be divided in three main groups with different dialects: the Riffians, the Chleuh and the Central Moroccan Amazigh. A large Jewish community lived in Morocco before the creation of Israel, numbering approximately 265,000 in 1948. Between 7,000 and 12,000 live there now, mostly in Casablanca, but also in Fes and other major cities. A call made by late king Hassan II to return to Morocco was not answered. Morocco is the world's third-largest producer of phosphorus and the price fluctuations of phosphates on the international market greatly influence Morocco's economy.

Name

The full Arabic name *al-Mamlaka al-Magribiyya* translates to "The Western Kingdom". *Al-Ma!rib* (meaning "The West") is commonly used. For historical references, medieval Arab historians and geographers used to refer to Morocco as *Al-Maghrib al Aqsa* ("The Farthest West"), disambiguating it from neighbouring historical regions called *al-Maghrib al Awsat* ("The Middle West", Algeria) and *al-Maghrib al Adna* ("The Nearest West", Tunisia).

The English name "Morocco" originates from Spanish "Marruecos" or the Portuguese "Marrocos", from medieval Latin "Morroch", which referred to the name of the former Almoravid and Almohad capital, Marrakesh. In Persian and Urdu and Hindi Morocco is still called "Marrakesh". In Turkish, Morocco is called "Fas" which comes from the ancient Idrisid and Marinid capital, Fes.

The word "Marrakesh" is probably derived from the Berber word combination *Mur-Akush*, meaning *Land of God.*

History

Berber Morocco

The area of present day Morocco has been inhabited since Neolithic times (at least since 8000 BC, as attested by signs of the Capsian culture), a period when the Maghreb was less arid than it is today. In Mesolithic ages the geography of Morocco resembled a savanna more than the present day arid landscape. In the classical period, Morocco was known as Mauretania, although this should not be confused with the modern-day country of Mauritania. Modern DNA analysis has confirmed that various populations have contributed to the present-day gene pool of Morocco in addition to the main ethnic group which is the Amazighs/Berbers. Those other various populations are Arabs, Iberians, Phoenicians, Sephardic Jews and sub-Saharan Africans.

Roman and pre-Roman Morocco

North Africa and Morocco were slowly drawn into the wider emerging Mediterranean world by Phoenician trading colonies and settlements in the early Classical period. Major early substantial settlements of the Phoenicians were at Chellah, Lixus and Mogador, with Mogador being a Phoenician colony as early as the early 6th century BC. The arrival of Phoenicians heralded a long engagement with the wider Mediterranean, as this strategic region formed part of the Roman Empire, as Mauretania Tingitana. In the fifth century, as the Roman Empire declined, the region fell to the Vandals, Visigoths, and then the Byzantine Empire, the Eastern Roman Empire, in rapid succession. During this time, however, the high mountains of most of modern Morocco remained unsubdued, and stayed in the hands of their Berber inhabitants. Christianity was introduced in the second century and gained converts in the towns and among slaves and Berber farmers.

Islamic Morocco

Islamic expansion began in the seventh century. In 670 AD, the first Islamic conquest of the North African coastal plain took place under Uqba ibn Nafi, a general serving under the Umayyads of Damascus. Arabs brought their customs, culture, and Islam, to which most of the Berbers converted, forming states and kingdoms such as the Kingdom of Nekor and Barghawata, sometimes after long-running series of civil wars. Under Idris ibn Abdallah who founded the Idrisid Dynasty, the country soon cut ties and broke away from the control of the distant Abbasid caliphs in Baghdad and the Umayyad rule in Al-Andalus. The Idrisids established Fes as their capital and Morocco became a centre of learning and a major regional power.

After the reign of the Idrisids, Arab settlers lost political control in the region of Morocco. After adopting Islam, Berber dynasties formed governments and reigned over the country. Morocco would reach its height under these Berber dynasties that replaced the Arab Idrisids after the 11th century. The Almoravids, the Almohads, then the Marinid and finally the Saadi dynasties would see Morocco rule most of Northwest Africa, as well as large sections of Islamic Iberia, or Al-Andalus. Following the Reconquista of the Iberian Peninsula, large numbers of Muslims and Jews fled to Morocco.

After the Saadi, the Arab Alaouite Dynasty eventually gained control. Morocco was facing aggression from Spain and the Ottoman Empire that was sweeping westward. The Alaouites succeeded in stabilizing their position, and while the kingdom was smaller than previous ones in the region, it remained quite wealthy. In 1684, they annexed Tangier. The organization of the kingdom developed under Ismail Ibn Sharif (1672–1727), who, against the opposition of local tribes began to create a unified state.

Morocco was the first nation to recognize the fledgling United States as an independent nation in 1787. In the beginning of the American Revolution, American merchant ships were subject to attack by the Barbary Pirates while sailing the Atlantic Ocean. On December 20, 1777, Morocco's Sultan Mohammed III declared that the American merchant ships would be under the protection of the sultanate and could thus enjoy safe passage. The Moroccan-American Treaty of Friendship stands as the U.S.'s oldest non-broken friendship treaty.

European Influence

Successful Portuguese efforts to invade and control the Atlantic coast in the fifteenth century did not profoundly affect the Mediterranean heart of Morocco. After the Napoleonic Wars, Egypt and the North African maghreb became increasingly ungovernable from Istanbul, the resort of pirates under local beys, and as Europe industrialized, an increasingly prized potential for colonization. The Maghreb had far greater proven wealth than the unknown rest of Africa and a location of strategic importance affecting the exit from the Mediterranean. For the first time, Morocco became a state of some interest in itself to the European Powers.

France showed a strong interest in Morocco as early as 1830. Recognition by the United Kingdom in 1904 of France's sphere of influence in Morocco provoked a reaction from the German Empire; the crisis of June 1905 was resolved at the Algeciras Conference, Spain in 1906, which formalized France's "special position" and entrusted policing of Morocco jointly to France and Spain. A second Moroccan crisis provoked by Berlin, increased tensions between European powers. The Treaty of Fez (signed on March 30, 1912) made Morocco a protectorate of France. By the same treaty, Spain assumed the role of protecting power over the northern and southern Saharan zones on November 27 that year.

Many Moroccan soldiers (Goumieres) served in the French army in both World War I and World War II, and in the Spanish Nationalist Army in the Spanish Civil War and after (*Regulares*).

Resistance

Nationalist political parties, which subsequently arose under the French protectorate, based their arguments for Moroccan independence on such World War II declarations as the Atlantic Charter (a joint U.S.-British statement that set forth, among other things, the right of all people to choose the form of government under which they live). A manifesto of the Istiqlal Party (Independence party in English) in 1944 was one of the earliest public demands for independence. That party subsequently provided most of the leadership for the nationalist movement.

France's exile of Sultan Mohammed V in 1953 to Madagascar and his replacement by the unpopular Mohammed Ben Aarafa, whose reign was perceived as illegitimate, sparked active opposition to the French and Spanish protectorates. The most notable violence occurred in Oujda where Moroccans attacked French and other European residents in the streets. Operations by the newly created "Jaish al-tahrir" (Liberation Army), were launched on October 1, 1955. Jaish al-tahrir was created by "Comite de Liberation du Maghreb Arabe" (Arab Maghreb Liberation Committee) in Cairo, Egypt to constitute a resistance movement against occupation. Its goal was the return of King Mohammed V and the liberation of Algeria and Tunisia as well. France allowed Mohammed V to return in 1955, and the negotiations that led to Moroccan independence began the following year.

All those events helped increase the degree of solidarity between the people and the newly returned king. For this reason, the revolution that Morocco knew was called "Taourat al-malik wa shaab" (The revolution of the King and the People) and it is celebrated every August 20.

Contemporary Morocco

On November 18, 2006, Morocco celebrated the 50th anniversary of its independence. Morocco recovered its political independence from France on March 2, 1956, and on April 7, France officially relinquished its protectorate. Through agreements with Spain in 1956 and 1958, Moroccan control over certain Spanish-ruled areas was restored, though attempts to claim other Spanish colonial possessions through military action were less successful.

The internationalized city of Tangier was reintegrated with the signing of the Tangier Protocol on October 29, 1956. Hassan II became King of Morocco on March 3, 1961. His early years of rule would be marked by political unrest. The Spanish enclave of Ifni in the south was reintegrated to the country in 1969. Morocco annexed the Western Sahara during the 1970s ("Marcha Verde", Green March) after demanding its reintegration from Spain since independence, but final resolution on the status of the territory remains unresolved.

Political reforms in the 1990s resulted in the establishment of a bicameral legislature in 1997. Morocco was granted Major non-NATO ally status by the United States in June 2004 and has signed free trade agreements with the United States and the European Union.

Politics

Morocco is a *de jure* constitutional monarchy with an elected parliament. The King of Morocco, with vast executive powers, can dissolve government and deploy the military, among other prerogatives. Opposition political parties are legal, and several have been formed in recent years.

Politics of Morocco take place in a framework of a parliamentary constitutional monarchy, whereby the Prime Minister of Morocco is the head of government, and of a multi-party system. Executive power is exercised by the government. Legislative power is vested in both the government and the two chambers of parliament, the Assembly of Representatives of Morocco and the Assembly of Councillors. The Moroccan Constitution provides for a monarchy with a Parliament and an independent judiciary.

The constitution grants the king extensive powers; he is both the secular political leader and the "Commander of the Faithful" as a direct descendant of the Prophet Mohammed. He presides over the Council of Ministers; appoints the Prime Minister following legislative elections, and on recommendations from the latter, appoints the members of the government. While the constitution theoretically allows the king to terminate the tenure of any minister, and after consultation with the heads of the higher and lower Assemblies, to dissolve the Parliament, suspend the constitution, call for new elections, or rule by decree, the only time this happened was in 1965. The King is formally the chief of the military. Upon the death of his father Mohammed V, King Hassan II succeeded to the throne in 1961. He ruled Morocco for the next 38 years, until he died in 1999. His son, King Mohammed VI, assumed the throne in July 1999.

Following the March 1998 elections, a coalition government headed by opposition socialist leader Abderrahmane Youssoufi and composed largely of ministers drawn from opposition parties, was formed. Prime Minister Youssoufi's government is the first government drawn primarily from opposition parties in decades, and also represents the first opportunity for a coalition of socialist, left-of-centre, and nationalist parties to be included in the government until October 2002. It was also the first time in the modern political history of the Arab world that the opposition assumed power following an election. The current government is headed by Abbas El Fassi.

Legislative Branch

Since the constitutional reform of 1996, the bicameral legislature consists of two chambers. The Assembly of Representatives of Morocco has 325 members elected for a five year term, 295 elected in multi-seat constituencies and 30 in national lists consisting only of women. The Assembly of Councillors (*Majlis al-Mustasharin*) has 270 members, elected for a nine year term, elected by local councils (162 seats), professional chambers (91 seats) and wage-earners (27 seats). The Parliament's powers, though limited, were expanded under the 1992 and 1996 constitutional revisions and include budgetary matters, approving bills, questioning ministers, and establishing ad hoc commissions of inquiry to investigate the government's actions. The lower chamber of Parliament may dissolve the government through a vote of no confidence.

Judicial Branch

The highest court in the judicial structure is the Supreme Court, whose judges are appointed by the King. The Youssoufi government continued to implement a reform program to develop greater judicial independence and impartiality. Morocco is divided into 16 administrative regions; the regions are administered by the Walis and governors appointed by the King.

Administrative Divisions

Morocco is divided into 16 regions, and subdivided into 62 prefectures and provinces. As part of a 1997 decentralization/regionalization law passed by the legislature, sixteen new regions were created. These regions are:

- Chaouia uardigha
- Doukkala-Abda
- Fes-Boulemane
- Gharb-Chrarda-Beni Hssen
- Grand Casablanca
- Guelmim-Es Semara
- Laayoune-Boujdour-Sakia El Hamra
- Marrakech-Tensift-El Haouz
- Meknes-Tafilalet
- Oriental
- Oued Ed-Dahab-Lagouira*
- Rabat-Sale-Zemmour-Zaer
- Souss-Massa-Draa
- Tadla-Azilal
- Tangier-Tetouan
- Taza-Al Hoceima-Taounate.

Provinces

Morocco is divided into 37 provinces and 2 wilayas*: Agadir, Al Hoceima, Azilal, Beni Mellal, Ben Slimane, Boulemane,

Casablanca*, Chaouen, El Jadida, El Kelaa des Sraghna, Er Rachidia, Essaouira, Fes, Figuig, Guelmim, Ifrane, Kenitra, Khemisset, Rommani, Khenifra, Khouribga, Laayoune, Larache, Marrakech, Meknes, Nador, Ouarzazate, Oujda, Rabat-Sale*, Safi, Settat, Sidi Kacem, Tangier, Tan-Tan, Taounate, Taroudannt, Tata, Taza, Tetouan, Tiznit; three additional provinces of Ad Dakhla (Oued Eddahab), Boujdour, and Es Smara as well as parts of Tan-Tan and Laayoune fall within Moroccan-claimed Western Sahara

Western Sahara Status

Because of the conflict over Western Sahara, the status of both regions of "Saguia el-Hamra" and "Rio de Oro" is disputed. The United Nations views Western Sahara as a non-self-governing territory, and as a case of unfinished decolonization. Morocco's rule in the territory is not internationally recognized, nor is the independent republic proposed by Polisario, a Saharawi group which fought against the Spanish colonial rule and then for Western Sahara's independence as the Sahrawi Arab Democratic Republic (today headquartered in Algeria). There is a ceasefire in effect since 1991, and a UN mission (MINURSO) is tasked with organizing a referendum on whether the territory should become independent or recognized as a part of Morocco. At the time, both parties signed an agreement to this effect, but Morocco has since 2000 refused to accept such a referendum, while Polisario demands that it be held.

The territory is mostly administered as the Southern Provinces by Morocco since Spain handed over the territory to Morocco and Mauritania after the Madrid Accords in 1975-76. Part of the territory, the Free Zone, is controlled by the Polisario Front as the Sahrawi Arab Democratic Republic. A UN-administered cease-fire has been in effect since September, 1991.

Western Sahara War

The Western Sahara War was the armed conflict which saw the Sahrawi rebel national liberation movement Polisario Front battling Spain, Morocco and Mauritania for the control of the former Spanish colony of Western Sahara from 1973 to 1991. The war resulted in the Spanish retreat in 1976, the

Mauritanian retreat in 1979 and a cease fire agreement with Morocco. The bigger part of the territory remained under Moroccan control.

Moroccan Autonomy Initiative

Recently, the government of Morocco has suggested autonomous status for the region, through the Moroccan Royal Advisory Council for Saharan Affairs (CORCAS). The project was presented to the United Nations Security Council in mid-April 2007. The proposal was encouraged by Moroccan allies such as the USA and France, but did not gain Security Council approval, and Minurso remains formally committed to the organization of a referendum. The Security Council has called upon the parties to enter into direct and unconditional negotiations to reach a mutually accepted political solution.

Geography

The geography of Morocco spans from the Atlantic Ocean, to mountainous areas, to the Sahara (desert). Morocco is a Northern African country, bordering the North Atlantic Ocean and the Mediterranean Sea, between Algeria and the annexed Western Sahara.

A large part of Morocco is mountainous. The Atlas Mountains are located mainly in the centre and the south of the country. The Rif Mountains are located in the north of the country. Both ranges are mainly inhabited by the Berber people. At 172,402 sq mi (446,519 km^2), Morocco is the fifty-seventh largest country in the world (after Uzbekistan). Algeria borders Morocco to the east and southeast though the border between the two countries has been closed since 1994.

There are also four Spanish enclaves on the Mediterranean coast: Ceuta, Melilla, Peñón de Vales de la Gomera, Peñón de Alhucemas, and the Chafarinas islands, as well as the disputed islet Perejil. Off the Atlantic coast the Canary Islands belong to Spain, whereas Madeira to the north is Portuguese. To the north, Morocco is bordered by and controls part of the Strait of Gibraltar, giving it power over the waterways in and out of the Mediterranean sea. The Rif mountains occupy the region bordering the Mediterranean from the north-west to the northeast. The Atlas Mountains run down the backbone of the country, from the south west to the north east. Most of the south east portion of the country is in the Sahara Desert and as such is generally sparsely populated and unproductive economically. Most of the population lives to the north of these mountains, while to the south is the desert. To the south, lies the Western Sahara, a former Spanish colony that was annexed by Morocco in 1975. Morocco claims that the Western Sahara is part of its territory and refers to that as its Southern Provinces.

Morocco's capital city is Rabat; its largest city is its main port, Casablanca. Other cities include Agadir, Essaouira, Fes, Marrakech, Meknes, Mohammadia, Oujda, Ouarzazat, Safi, Sale, Tangier and Tetouan.

Morocco is represented in the ISO 3166-1 alpha-2 geographical encoding standard by the symbol *MA*. This code was used as the basis for Morocco's internet domain,.ma.

Climate

The climate is Mediterranean in the North and in some mountains (West of Atlas), which becomes more extreme towards the interior regions. The terrain is such that the coastal plains are rich and accordingly, they comprise the backbone for agriculture, especially in the North. Forests cover about 12% of the land while arable land accounts for 18%. 5% is irrigated. In the Atlas (Middle Atlas), there are several different climates: Mediterranean (with some more humid and fresher variants), Maritime Temperate (with some humid and fresher variants too) that allow different species of oaks, moss carpets, junipers, atlantic cedars and many other plants, to form extensive and very rich humid cloud forests. In the highest peaks a different climate may occur. On the other side of Atlas mountains (East Atlas), the climate changes, due to the barrier/shelter effect of these mountainous system, turning it very dry and extremely warm during the summer (that can last several months), especially on the lowlands and on the valleys faced to the Sahara. Here it starts the big Desert Sahara and it's perfectly visible, for example, on the Draa Valley, on which it's possible to find oases, sand dunes and rocky desert landscapes. So the climate in this region is desert.

Wildlife

Morocco is known for its wildlife biodiversity. Birds represent the most important fauna. The avifauna of Morocco includes a total of 454 species, of which five have been introduced by humans, and 156 are rare or accidental.

Economy

Morocco's economy is considered a relatively liberal economy governed by the law of supply and demand. Since 1993, the country has followed a policy of privatization of certain economic sectors which used to be in the hands of the government.

Government reforms and steady yearly growth in the region of 4-5% from 2000 to 2007, including 4.9% year-on-year growth in 2003-2007 helped the Moroccan economy to become much more robust compared to a few years ago. Economic growth is far more diversified, with new service and industrial poles, like Casablanca and Tangier, developing. The agriculture sector is being rehabilitated, which in combination with good rainfalls led to a growth of over 20% in 2009.

The services sector accounts for just over half of GDP and industry, made up of mining, construction and manufacturing, is an additional quarter. The sectors who recorded the highest growth are the tourism, telecoms, information technology, and textile sectors. Morocco, however, still depends to an inordinate degree on agriculture. The sector accounts for only around 14% of GDP but employs 40-45% of the Moroccan population. With a semi-arid climate, it is difficult to assure good rainfall and Morocco's GDP varies depending on the weather. Fiscal prudence has allowed for consolidation, with both the budget deficit and debt falling as a percentage of GDP.

The economic system of the country presents several facets. It is characterized by a large opening towards the outside world. France remains the primary trade partner (supplier and customer) of Morocco. France is also the primary creditor and foreign investor in Morocco. In the Arab world, Morocco has the second-largest non-oil GDP, behind Egypt, as of 2005.

Since the early 1980s the Moroccan government has pursued an economic program toward accelerating real economy growth with the support of the International Monetary Fund, the

World Bank, and the Paris Club of creditors. The country's currency, the dirham, is now fully convertible for current account transactions; reforms of the financial sector have been implemented; and state enterprises are being privatized.

The major resources of the Moroccan economy are agriculture, phosphates, and tourism. Sales of fish and seafood are important as well. Industry and mining contribute about one-third of the annual GDP. Morocco is the world's third-largest producer of phosphorus (after China, which is first, and the United States which is second), and the price fluctuations of phosphates on the international market greatly influence Morocco's economy. Tourism and workers' remittances have played a critical role since the Kingdom's independence. The production of textiles and clothing is part of a growing manufacturing sector that accounted for approximately 34% of total exports in 2002, employing 40% of the industrial workforce. The government wishes to increase textile and clothing exports from $1.27 billion in 2001 to $3.29 billion in 2010.

The information technology and communications sectors have been witnessing significant expansion as well. Morocco is the first country in North Africa to install a 3G network. The number of Internet subscribers in the country jumped 73% in 2006 over the year-ago period. Further, a new offshore site at Casablanca, with state-of-the-art technologies and other incentives, has grabbed the attention of many global multinationals. Setting up offshore service centres in the nation has become tempting. Such is the rate of growth, that off-shoring and IT activities are estimated to contribute $500 million to the country's GDP and employ 30,000 people by 2015. The communications sector already accounts for half of all foreign direct investments Morocco received over the past five years.

The high cost of imports, especially of petroleum imports, is a major problem. Another chronic problem is unreliable rainfall, which produces drought or sudden floods; in 1995, the country's worst drought in 30 years forced Morocco to import grain and adversely affected the economy. Another drought occurred in 1997, and one in 1999–2000. Reduced incomes due to drought caused GDP to fall by 7.6% in 1995, by 2.3% in 1997, and by 1.5% in 1999. During the years between drought, good rains brought bumper crops to market. Good rainfall in 2001 led to a 5% GDP growth rate. Morocco suffers both from unemployment (9.6% in 2008), and a large external debt estimated at around $20 billion, or half of GDP in 2002.

Among the various free trade agreements that Morocco has ratified with its principal economic partners, are The Euro-Mediterranean free trade area agreement with the European Union with the objective of integrating the European Free Trade Association at the horizons of 2012; the Agadir Agreement, signed with Egypt, Jordan, and Tunisia, within the framework of the installation of the Greater Arab Free Trade Area; the US-Morocco Free Trade Agreement with United States which came into force in January 1, 2006, and lately the agreement of free exchange with Turkey.

Demographics

Morocco is the fourth most populous Arab country, after Egypt, Sudan and Algeria. Most Moroccans practice Sunni Islam and are of Berber, Arab or mixed Arab-Berber stock. Arabs and Berbers comprise about 99.1% of the Moroccan population.

Morocco has been inhabited by Berbers for at least the last 5000 years. The Arabs conquered the territory that would become Morocco in the 7th and 11th centuries, at the time under the rule of various late Byzantine Roman leaders and indigenous Berber and Romano-Berber principalities, laying the foundation for the emergence of an Arab-Berber culture. A sizeable portion of the population is identified as Haratin and Gnawa (or Gnaoua), black or mixed race. Morocco's Jewish minority (265,000 in 1948) has decreased significantly and numbers about 5,500. Most of the 100,000 foreign residents are French or Spanish, largely colonists' descendants, who primarily work for European multinational companies. Prior to independence, Morocco was home to half a million Europeans, mainly Spanish and French settlers (*colons*).

Recent studies make clear no significant genetic differences exist between Arabic and non-Arabic speaking populations, highlighting that in common with most of the Arab World,

Arabization was mainly via acculturation of indigenous populations over time. According to the *European Journal of Human Genetics*, Moroccans from North-Western Africa were genetically closer to Iberians than to Sub-Saharan Africans of Bantu ethnicity.

The largest concentration of Moroccans outside Morocco is in France, which has reportedly over one million Moroccans. There are also large Moroccan communities in Spain (about 700,000 Moroccans), the Netherlands, Belgium, Italy and Canada.

Most people live west of the Atlas Mountains, a range that insulates the country from the Sahara Desert. Casablanca is the centre of commerce and industry and the leading port; Rabat is the seat of government; Tangier is the gateway to Morocco from Spain and also a major port; Fes is the cultural and religious centre; and Marrakech is a major tourist centre.

There is a European expatriate population of 100,000, mainly of French or Spanish descent; many are teachers or technicians and more and more retirees, especially in Marrakech.

Languages

Morocco's official language is Modern Standard Arabic. The country's distinctive Arabic dialect is called Moroccan Arabic. Approximately 10 million (30% of the population), mostly in rural areas, speak Berber – which exists in Morocco in three different dialects (Tarifit, Tashelhiyt, and Tamazight) – either as a first language or bilingually with the spoken Arabic dialect. French, which is Morocco's unofficial second language, is taught universally and serves as Morocco's primary language of commerce and economics. It also is widely used in education and government. About 2 million Moroccans in the northern part of the country speak Spanish as a second language in parallel with Tarifit. English, while still far behind French and Spanish in terms of number of speakers, is rapidly becoming the second foreign language of choice among educated youth. As a result of national education reforms entering into force in late 2002, English will be taught in all public schools from the fourth year on. French however, will remain the second language because of Morocco's close economic and social links with other French-speaking countries and especially France.

Berber

The number of Tarifit dialect was estimated to have around 1.5 million in 1990. The language is spoken in the Rif area in the north of the country, and is the smallest Berber dialect in Morocco, by number of speakers. The Tashelhiyt dialect is considered to be the most widely spoken Berber dialect, as it covers the whole of the Region Souss-Massa-Draa, and is also spoken in the Marrakech-Tensift-El Haouz region. Studies done in 1990 show around 3 million people, concentrated in the south of Morocco, speak the dialect.

Linguistically, Berber belongs to the Afro-Asiatic group, and has many accents or variants. The three main accents used in Morocco are Tachelhit, Tamazight and Tarifit (Also called *Thamazight* by its speakers). Collectively, those Berber languages they are known as "Chelha" in Moroccan Arabic and as "Barbaria" in Classical Arabic used in the Middle East. The terms "Barbar" and "Chelha" are considered by most Berber activists as extremely offending and humiliating. They prefer the word *Amazigh*.

Culture

Morocco is an ethnically diverse country with a rich culture and civilization. Through Moroccan history, Morocco hosted many people coming from East (Phoenicians, Carthaginians, Jews and Arabs), South (Sub-Saharan Africans) and North (Romans, Vandals, Andalusians (including Moors and Jews). All those civilizations have had an impact on the social structure of Morocco. It conceived various forms of beliefs, from paganism, Judaism, and Christianity to Islam.

The production of Moroccan literature has continued to grow and diversify. To the traditional genres—poetry, essays, and historiography—have been added forms inspired by Middle Eastern and Western literary models. French is often used in publishing research in the social and natural sciences, and in the fields of literature and literary studies, works are published in both Arabic and French. Moroccan writers, such as Mohammed Choukri, Driss Chraïbi, Abdallah Laroui,

Abdelfattah Kilito, and Fatima Mernissi, publish their works in both French and English. Expatriate writers such as Pierre Loti, William S. Burroughs, and Paul Bowles have drawn attention to Moroccan writers as well as to the country itself.

Since independence a veritable blossoming has taken place in painting and sculpture, popular music, amateur theatre, and filmmaking. The Moroccan National Theatre (founded 1956) offers regular productions of Moroccan and French dramatic works. Art and music festivals take place throughout the country during the summer months, among them the World Sacred Music Festival at Fes.

Moroccan music, influenced by Arab, Amazigh, African, and Andalusian traditions, makes use of a number of traditional instruments, such as the flute (nay), shawm (ghaita), zither (qanun), and various short necked lutes. These are often backed by explosive percussion on the darbukka (terra-cotta drum). Among the most popular traditional Moroccan artists internationally are the Master Musicians of Jajouka, an all-male guild trained from childhood, and Hassan Hakmoun, a master of gnawa trance music, a popular spiritual style that traces its roots to sub-Saharan Africa. Younger Moroccans enjoy rai, a style of plain-speaking Algerian music that incorporates traditional sounds with those of Western rock, Jamaican reggae, and Egyptian and Moroccan popular music.

Each region possesses its own specificities, thus contributing to the national culture and to the legacy of civilization. Morocco has set among its top priorities the protection of its diverse legacy and the preservation of its cultural heritage.

Culturally speaking, Morocco has always been successful in combining its Berber, Jewish and Arabic cultural heritage with external influences such as the French and the Spanish and, during the last decades, the Anglo-American lifestyles.

Cuisine

Moroccan cuisine has long been considered as one of the most diversified cuisines in the world. This is a result of the centuries-long interaction of Morocco with the outside world. The cuisine of Morocco is a mix of Berber, Spanish, Corsican, Portuguese, Moorish, Middle Eastern, Mediterranean, and

African cuisines. The cuisine of Morocco has been influenced by the native Berber cuisine, the Arabic Andalusian cuisine brought by the Moriscos when they left Spain, the Turkish cuisine from the Turks and the Middle Eastern cuisines brought by the Arabs, as well as Jewish cuisine.

Spices are used extensively in Moroccan food. While spices have been imported to Morocco for thousands of years, many ingredients, like saffron from Tiliouine, mint and olives from Meknes, and oranges and lemons from Fez, are home-grown. Chicken is the most widely eaten meat in Morocco. The most commonly eaten red meat in Morocco is beef; lamb is preferred but is relatively expensive. Couscous is the most famous Moroccan dish along with pastilla, tajine, and harira. The most popular drink is green tea with mint.

Literature

Moroccan literature is written in Arabic, Berber and French. It also contains literature produced in Al-Andalus. Under the Almohad dynasty Morocco experienced a period of prosperity and brilliance of learning. The Almohad built the Marrakech Koutoubia Mosque, which accommodated no fewer than 25,000 people, but was also famed for its books, manuscripts, libraries and book shops, which gave it its name; the first book bazaar in history. The Almohad Caliph Abu Yakub had a great love for collecting books. He founded a great library, which was eventually carried to the Casbah and turned into a public library.

Modern Moroccan literature began in the 1930s. Two main factors gave Morocco a pulse toward witnessing the birth of a modern literature. Morocco, as a French and Spanish protectorate left Moroccan intellectuals the opportunity to exchange and to produce literary works freely enjoying the contact of other Arabic literature and Europe.

During the 1950s and 1960s, Morocco was a refuge and artistic centre and attracted writers as Paul Bowles, Tennessee Williams and William S. Burroughs. Moroccan literature flourished with novelists such as Mohamed Zafzaf and Mohamed Choukri, who wrote in Arabic, and Driss Chraïbi and Tahar Ben Jelloun who wrote in French. Other important Moroccan authors include, Abdellatif Laabi, Abdelkarim Ghellab, Fouad Laroui, Mohammed Berrada and Leila Abouzeid. It should be noted also, that orature (oral literature) is an integral part of Moroccan culture, be it in Moroccan Arabic or Amazigh.

Music

Moroccan music is of Arab, Berber and sub-saharan origins. Rock-influenced chaabi bands are widespread, as is trance music with historical origins in Muslim music.

Morocco is home to Andalusian classical music that is found throughout North Africa. It probably evolved under the Moors in Cordoba, and the Persian-born musician Ziryab is usually credited with its invention. A genre known as Contemporary Andalusian music and art is the brainchild of Morisco visual artist/composer/oudist Tarik Banzi founder of the Al-Andalus Ensemble

Chaabi (*popular*) is a music consisting of numerous varieties which are descended from the multifarious forms of Moroccan folk music. Chaabi was originally performed in markets, but is now found at any celebration or meeting.

Popular Western forms of music are becoming increasingly popular in Morocco, such as fusion, rock, country, metal and particularly hip hop.

Transport

The railway network of Morocco consists of 1907 km 1,435 mm (4 ft 8+1D 2 in) standard gauge and 1003 km electrified with 3 kV DC. There are connections to Algeria, and consecutively Tunisia, but since the 1990s the connections are closed. The Gibraltar Tunnel is a rail tunnel link proposed between Tangier, Morocco and Spain under the Strait of Gibraltar to be in operation in 2025.

There are plans for high-speed lines: Work by ONCF could begin in 2007 from Marrakech to Tangier in the north via Marrakesh to Agadir in the south, and from Casablanca on the Atlantic to Oujda on the Algerian border. If the plans are approved, the 1,500 kilometres of track may take until 2030 to complete at a cost of around 25 billion dirhams ($3.37 billion). Casablanca to Marrakesh could be cut to 1 hour and 20 minutes from over three hours, and from the capital Rabat to Tangier to 1 hour and 30 minutes from 4 hours and 30 minutes.

There are around 56986 kilometres of roads (national, regional and provincial) in Morocco. In addition to 610,5 kilometre of highways.

Military

Compulsory military service in Morocco has been suppressed since September 2006, and the country's reserve obligation lasts until age 50. The country's military consists of the Royal Armed Forces—this includes the army (the largest branch) and a small navy and air force—the National Police Force, the Royal Gendarmerie (mainly responsible for rural security), and the Auxiliary Forces. Internal security is generally effective, and acts of political violence are rare (with one exception, the 2003 Casablanca bombings which killed 45 people). The UN maintains a small observer force in Western Sahara, where a large number of Morocco's troops are stationed. The Saharawi group Polisario maintains an active militia of an estimated 5,000 fighters in Western Sahara and has engaged in intermittent warfare with Moroccan forces since the 1980s.

The military of Morocco is composed of the following main divisions:

- Royal Armed Forces
  - Army
  - Navy
  - Air Force
  - Gendarmerie
  - Auxiliary Forces
  - Moroccan Royal Guard
  - Marche Verte.

Education

Education in Morocco is free and compulsory through primary school (age 15). Nevertheless, many children – particularly girls in rural areas – still do not attend school. The country's illiteracy rate has been stuck at around 50% for some years, but reaches as high as 90% among girls in rural regions.

On September 2006, UNESCO awarded Morocco amongst other countries such as Cuba, Pakistan, Rajasthan (India) and Turkey the "UNESCO 2006 Literacy Prize".

Morocco has about 230,000 students enrolled in fourteen public universities. The Mohammed V University in Rabat and Al Akhawayn University in Ifrane (a private university) are highly regarded. Al-Akhawayn, founded in 1993 by King Hassan II and King Fahd of Saudi Arabia, is an English-language American-style university comprising about 1,000 students. The University of Al Karaouine, in Fez, is considered the oldest continuously operating university in the world and has been a centre of learning for more than 1,000 years.

Morocco allocates approximately one fifth of its budget to education. Much of this is spent on building schools to accommodate the rapidly growing population. Education is mandatory for children between the ages of 7 and 13 years. In urban areas the majority of children in this age group attend school, though on a national scale the level of participation drops significantly. About three fourths of school age males attend school, but only about half of school age girls; these proportions drop markedly in rural areas. Slightly more than half of the children go on to secondary education, including trade and technical schools. Of these, few seek higher education. Poor school attendance, particularly in rural areas, has meant a low rate of literacy, which is about two fifths of the population.

Universities

Morocco has more than four dozen universities, institutes of higher learning, and polytechnics dispersed at urban centres throughout the country. Its leading institutions include Muhammad V University in Rabat, the country's largest university, with branches in Casablanca and Fes; the Hassan II Agriculture and Veterinary Institute in Rabat, which conducts leading social science research in addition to its agricultural specialties; and Al-Akhawayn University in Ifrane, the first English-language university in North Africa, inaugurated in 1995 with contributions from Saudi Arabia and the United States. The University of Al-Karaouine or Al-Qarawiyyin is a university located in Fes. It is considered the oldest continuously operating academic degree-granting university in the world.

Sport

Spectator sports in Morocco traditionally centred on the art of horsemanship until European sports—football (soccer), polo, swimming, and tennis—were introduced at the end of the 19th century. Football is the country's premier sport, popular among the urban youth in particular, and in 1970 Morocco became the first African country to play in World Cup competition. At the 1984 Olympic Games, two Moroccans won gold medals in track and field events, one of whom—Nawal El Moutawakel in the 400 metre hurdles—was the first woman from an Arab or Islamic country to win an Olympic gold medal. Another was Hicham El Guerrouj. Tennis and golf have also become popular. Several Moroccan professional players have competed in international competition, and the country fielded its first Davis Cup team in 1999.

As of 2007, Moroccan society participated in many sports, including handball, football, golf, tennis, basketball, and athletics. Hicham El Guerrouj, a retired middle distance runner for Morocco, won 2 gold medals for Morocco at the Athletics at the 2004 Summer Olympics and holds the 1.609 km (1 mile) world record, along with other notable performances.

6

The Republic of the Niger

Niger, officially named the Republic of Niger, is a landlocked country in Western Africa, named after the Niger River. It borders Nigeria and Benin to the south, Burkina Faso and Mali to the west, Algeria and Libya to the north and Chad to the east. Niger covers a land area of almost 1,270,000 km, over 80 percent of which is covered by the Sahara desert. The country's predominantly Islamic population of just above 15,000,000 is mostly clustered in the far south and west of the nation. The capital city is Niamey.

Niger is a developing country. Much of the non-desert portions of the country are threatened by periodic drought and desertification.

The economy is concentrated around subsistence and some export agriculture clustered in the more fertile south, and the export of raw materials—especially uranium ore. Niger remains handicapped by its landlocked position, desert terrain, poor education and poverty of its people, lack of infrastructure, poor health care, and environmental degradation.

Nigerien society reflects a great diversity drawn from the long independent histories of its several ethnic groups and regions and their relatively short period living in a single state. Historically, what is now Niger has been on the fringes of several large states. Since independence, Nigeriens have lived under five constitutions and three periods of military rule. A majority live in rural areas, and have little access to advanced education.

Geography

Niger is a landlocked nation in West Africa located along the border between the Sahara and Sub-Saharan regions. Its geographic coordinates are latitude 16°N and longitude 8°E. Its area is 1,267,000 square kilometres (489,191 sq mi) of which 300 square kilometres (116 sq mi) is water. This makes Niger slightly less than twice the size of the U.S. state of Texas, and the world's twenty-second largest country (after Chad). Niger is comparable in size to Angola.

Niger borders seven countries on all sides and has a total of 5,697 kilometres (3,540 mi) of borders. The longest border is with Nigeria to the south (1,497 km/930 mi). This is followed by Chad to the east, at 1,175 km (730 mi), Algeria to the north-northwest (956 km/594 mi), and Mali at 821 km (510 mi). Niger also has small borders in its far southwest frontier with Burkina Faso at 628 km (390 mi) and Benin at 266 km (165 mi) and to the north-northeast (Libya at 354 km (220 mi).

Climate

Niger's subtropical climate is mainly very hot and dry, with much desert area. In the extreme south there is a tropical climate on the edges of the Niger River basin. The terrain is predominantly desert plains and sand dunes, with flat to rolling savanna in the south and hills in the north.

History

While most of what is now Niger has been subsumed into the inhospitable Sahara desert in the last two thousand years, five thousand years ago the north of the country was fertile grasslands. Populations of pastoralists have left paintings of abundant wildlife, domesticated animals, chariots, and a complex culture that dates back to at least 10,000 BCE. Several former northern villages and archaeological sites date from the Green Sahara period of 7,500-7,000 to 3,500-3,000BCE

Early Historical Period

The Songhai Empire expanded into what is modern Niger from the 1400s, reaching as far as Agadez before its collapse in 1591, from which the modern Zarma and Songhai peoples trace their history. At its fall, portions of the empire and refugees from modern Mali formed a series of Songhai states, with the Dendi Kingdom becoming the most powerful. From the 1200s, the nomadic Tuareg formed large confederations, pushed southward, into the Air Mountains, displacing some previous residents to the south. At their peak, the Tuareg confederations ruled most of what is now northern Niger, and extended their influence into modern Nigeria.

In the 18th century, Fula pastoralists moved into the Liptako area of the west, while smaller Zarma kingdoms, siding with various Hausa states, clashed with the expanding Fulani Empire of Sokoto from the south. The colonial border with British Nigeria was in part based on the rupture between the Sokoto Caliphate to the south, and Hausa ruling dynasties which had fled to the north. In the far east around the Lake Chad basin, the successive expansion of the Kanem Empire and Bornu Empire spread ethnically Kanuri and Toubou rulers and their subject states as far west as Zinder and the Kaouar Oases from the 10th to the 17th centuries.

In the 19th century, contact with the West began when the first European explorers—notably Mungo Park (British) and Heinrich Barth (German)—explored the area, searching for the source of the Niger River. Although French efforts at "pacification" began before 1900, dissident ethnic groups, especially the desert Tuareg, were not fully subdued until 1922, when Niger became a French colony.

Niger's colonial history and development parallel that of other French West African territories. France administered its West African colonies through a governor general in Dakar, Senegal, and governors in the individual territories, including Niger. In addition to conferring French citizenship on the inhabitants of the territories, the 1946 French constitution provided for decentralization of power and limited participation in political life for local advisory assemblies.

Early Independence

A further revision in the organization of overseas territories occurred with the passage of the Overseas Reform Act (*Loi Cadre*) of July 23, 1956, followed by reorganizing measures enacted by the French Parliament early in 1957. In addition to removing voting inequalities, these laws provided for creation of governmental organs, assuring individual territories a large measure of self-government. After the establishment of the Fifth French Republic on December 4, 1958, Niger became an autonomous state within the French Community. Following full independence on August 3, 1960, however, membership was allowed to lapse.

Single Party and Military Rule (1961-1991)

For its first fourteen years as an independent state, Niger was run by a single-party civilian regime under the presidency of Hamani Diori. In 1974, a combination of devastating drought and accusations of rampant corruption resulted in a coup d'etat that overthrew the Diori regime. Col. Seyni Kountche and a small military group ruled the country until Kountche's death in 1987.

He was succeeded by his Chief of Staff, Col. Ali Saibou, who released political prisoners, liberalized some of Niger's laws and policies, and promulgated a new constitution, with the creation of a single party constitutional Second Republic. However, President Saibou's efforts to control political reforms failed in the face of union and student demands to institute a multi-party democratic system. The Saibou regime acquiesced to these demands by the end of 1990.

New political parties and civic associations sprang up, and a national peace conference was convened in July 1991 to prepare the way for the adoption of a new constitution and the holding of free and fair elections. The debate was often contentious and accusatory, but under the leadership of Prof. Andre Salifou, the conference developed a plan for a transition government.

Third Republic

This caretaker government was installed in November 1991 to manage the affairs of state until the institutions of the Third Republic were put into place in April 1993. While the economy deteriorated over the course of the transition, certain accomplishments stand out, including the successful conduct of a constitutional referendum; the adoption of key legislation such as the electoral and rural codes; and the holding of several free, fair, and non-violent nationwide elections. Freedom of the press flourished with the appearance of several new independent newspapers.

The results of the January 1995 parliamentary election meant cohabitation between a rival president and prime minister; this led to governmental paralysis, which provided Col. Ibrahim Bare Mainassara a rationale to overthrow the Third Republic in January 1996.

Military Rule and the Fourth Republic

While leading a military authority that ran the government (Conseil de Salut National) during a 6-month transition period, Bare enlisted specialists to draft a new constitution for a Fourth Republic announced in May 1996. Bare organized a presidential election in July 1996. While voting was still going on, he replaced the electoral commission. The new commission declared him the winner after the polls closed. His party won 57% of parliament seats in a flawed legislative election in November 1996. When his efforts to justify his coup and subsequent questionable elections failed to convince donors to restore multilateral and bilateral economic assistance, a desperate Bare ignored an international embargo against Libya and sought Libyan funds to aid Niger's economy. In repeated violations of basic civil liberties by the regime, opposition leaders were imprisoned; journalists often arrested, and deported by an unofficial militia composed of police and military; and independent media offices were looted and burned.

As part of an initiative started under the 1991 national conference, however, the government signed peace accords in April 1995 with all, meaning Tuareg and Toubou groups that had been in rebellion since 1990. The Tuareg claimed they lacked attention and resources from the central government. The government agreed to absorb some former rebels into the military and, with French assistance, help others return to a productive civilian life.

Fifth Republic Since 1999

On April 9, 1999, Bare was killed in a coup led by Maj. Daouda Malam Wanke, who established a transitional National

Reconciliation Council to oversee the drafting of a constitution for a Fifth Republic with a French style semi-presidential system.

In votes that international observers found to be generally free and fair, the Nigerien electorate approved the new constitution in July 1999 and held legislative and presidential elections in October and November 1999. Heading a coalition of the National Movement for a Developing Society (MNSD) and the Democratic and Social Convention (CDS), Mamadou Tandja won the election.

In a February 2010 coup d'etat, a military junta was established in response to Tandja's attempted extension of his political term through constitutional manipulation. The coup established a junta led by the Supreme Council for the Restoration of Democracy.

Politics

Niger's new constitution was approved in July 1999. It restored the semi-presidential system of government of the December 1992 constitution (Third Republic) in which the president of the republic, elected by universal suffrage for a five-year term, and a prime minister named by the president share executive power. As a reflection of Niger's increasing population, the unicameral National Assembly was expanded in 2004 to 113 deputies elected for a 5 year term under a majority system of representation. Political parties must attain at least 5% of the vote in order to gain a seat in the legislature.

The constitution also provides for the popular election of municipal and local officials, and the first-ever successful municipal elections took place on July 24, 2004. The National Assembly passed in June 2002 a series of decentralization bills. As a first step, administrative powers will be distributed among 265 communes (local councils); in later stages, regions and departments will be established as decentralized entities. A new electoral code was adopted to reflect the decentralization context. The country is currently divided into 8 regions, which are subdivided into 36 districts (departments). The chief administrator (Governor) in each department is appointed by the government and functions primarily as the local agent of the central authorities. The current legislature elected in December 2004 contains seven political parties. President Mamadou Tandja was re-elected in December 2004 and reappointed Hama Amadou as Prime Minister. Mahamane Ousmane, the head of the CDS, was re-elected President of the National Assembly (parliament) by his peers. The new second term government of the Fifth Republic took office on December 30, 2002. In August 2002, serious unrest within the military occurred in Niamey, Diffa, and Nguigmi, but the government was able to restore order within several days.

In June 2007, Seyni Oumarou was nominated as the new Prime Minister after Hama Amadou was democratically forced out of office by the National Assembly through a motion of no confidence.

From 2007 to 2008, the Second Tuareg Rebellion took place in northern Niger, worsening economic prospects and shutting down political progress.

On 26 May 2009, President Tandja dissolved parliament after the country's constitutional court ruled against plans to hold a referendum on whether to allow him a third term in office. According to the constitution, a new parliament was elected within three months. This touched off a political struggle between Tandja, trying to extend his term-limited authority beyond 2009 through the establishment of a Sixth Republic, and his opponents who demanded that he step down at the end of his second term in December 2009. See 2009 Nigerien constitutional crisis.

Regions, Departments and Communes

Niger is divided into 7 Regions and one capital district. These Regions are subdivided into 36 departments. The 36 Departments are currently broken down into Communes of varying types. As of 2006 there were 265 communes, including communes urbaines (Urban Communes: as subdivisions of major cities), communes rurales (Rural Communes, in sparsely populated areas and postes administratifs (Administrative Posts) for largely uninhabited desert areas or military zones.

Rural communes may contain official villages and settlements, while Urban Communes are divided into quarters.

Niger subvisions were renamed in 2002, in the implementation of a decentralisation project, first begun in 1998. Previously, Niger was divided into 7 Departments, 36 Arrondissements, and Communes.

These subdivisions were administered by officials appointed by the national government. These offices will be replaced in the future by democratically elected councils at each level.

The departments and capital district are:

- Agadez Region
- Diffa Region
- Dosso Region
- Maradi Region
- Tahoua Region
- Tillaberi Region
- Zinder Region
- Niamey (capital district).

Foreign Relations

Niger pursues a moderate foreign policy and maintains friendly relations with the West and the Islamic world as well as nonaligned countries. It belongs to the United Nations and its main specialized agencies and in 1980-81 served on the UN Security Council. Niger maintains a special relationship with former colonial power France and enjoys close relations with its West African neighbours.

It is a charter member of the African Union and the West African Monetary Union and also belongs to the Niger Basin Authority and Lake Chad Basin Commission, the Economic Community of West African States, the Non-Aligned Movement, the Organization of the Islamic Conference and the Organization for the Harmonization of Business Law in Africa (OHADA). The western most regions of Niger are joined with contiguous regions Mali and Burkina Faso under the Liptako-Gourma Authority.

The border dispute with Benin, inherited from colonial times and concerning inter alia Lete Island in the River Niger was finally solved by the ICJ in 2005 to Niger's advantage.

Military

The Niger Armed Forces total 12,000 personnel with approximately 3,700 gendarmes, 300 air force, and 6,000 army personnel. The air force has four operational transport aircraft. The armed forces include general staff and battalion task force organizations consisting of two paratroop units, four light armored units, and nine motorized infantry units located in Tahoua, Agadez, Dirkou, Zinder, Nguigmi, N'Gourti, and Madewela. Since January 2003, Niger has deployed a company of troops to Cote d'Ivoire as part of the ECOWAS stabilization force. In 1991, Niger sent four hundred military personnel to join the American-led allied forces against Iraq during the Gulf War.

Niger's defence budget is modest, accounting for about 1.6% of government expenditures. France provides the largest share of military assistance to Niger. Morocco, Algeria, China, and Libya have also provided military assistance. Approximately 15 French military advisers are in Niger. Many Nigerien military personnel receive training in France, and the Nigerien Armed Forces are equipped mainly with military hardware either sold or donated by France.

In the past, U.S. assistance focused on training pilots and aviation support personnel, professional military education for staff officers, and initial specialty training for junior officers. A small foreign military assistance program was initiated in 1983. A U.S. Defence Attache office opened in June 1985 and assumed Security Assistance Office responsibilities in 1987. The office closed in 1996 following a coup d'etat. A U.S. Defence Attache office reopened in July 2000. The United States provided transportation and logistical assistance to Nigerien troops deployed to Cote d'Ivoire in 2003. Additionally, the U.S. provided initial equipment training on vehicles and communications gear to a select contingent of Nigerien soldiers as part of the Department of State Pan Sahel Initiative.

In February 2010, the army of Niger staged another coup d'etat, that ousted President Tandja Mamadou, who had been behaving in an increasingly dictatorial fashion. The army claims to be acting toward the restoration of democracy. It is still unclear whether this will actually happen.

Transport

Transport is crucial to the economy and culture of this vast landlocked nation, with cities separated by huge uninhabited deserts, mountain ranges, and other natural features. Niger's transport system was little developed during the colonial period (1899–1960), relying upon animal transport, human transport, and limited river transport in the far south west and south east.

No railways were constructed in the colonial period, and most roads outside the capital remained unpaved. The Niger River is unsuitable for river transport of any large scale, as it lacks depth for most of the year, and is broken by rapids at many spots. Camel caravan transport was historically important in the Sahara desert and Sahel regions which cover most of the north.

Road

Road transport, especially shared taxis, buses, and trucks, are the primary form of long distance transport for most Nigeriens. There were 10,100 km of roads in the nation in 1996, but only 798 km were paved. Most of this total was in large cities and in two main highways.

The first major paved highway was constructed in the 1970s and 80s to transport uranium from the far northern mining town of Arlit to the Benin border. (Much of Niger's export economy relies upon ports in Cotonou, Lome, and Port Harcourt.) This road, dubbed the *Uranium Highway* runs through Arlit, Agadez, Tahoua, Birnin-Konni, and Niamey, and is part of the Trans-Sahara Highway system. The paved RN1 ("Routes National") runs east-west across the south of the nation, from Niamey via Maradi and Zinder towards Diffa in the far east of the nation, although the stretch from Zinder to Diffa is only partially paved. Other roads range from all-weather laterite surfaces to grated dirt or sand pistes, especially in the desert north. These form a more extensive numbered highway system.

Air Transport

Niger's main international airport is Diori Hamani International Airport at Niamey. Other airports in Niger include

Mano Dayak International Airport at Agadez and Zinder Airport near Zinder.

Economy

The economy of Niger centres on subsistence crops, livestock, and some of the world's largest uranium deposits. Drought cycles, desertification, a 2.9% population growth rate, and the drop in world demand for uranium have undercut the economy.

Niger shares a common currency, the CFA franc, and a common central bank, the Central Bank of West African States (BCEAO), with seven other members of the West African Monetary Union. Niger is also a member of the Organization for the Harmonization of Business Law in Africa (OHADA).

In December 2000, Niger qualified for enhanced debt relief under the International Monetary Fund program for Heavily Indebted Poor Countries (HIPC) and concluded an agreement with the Fund for Poverty Reduction and Growth Facility (PRGF). Debt relief provided under the enhanced HIPC initiative significantly reduces Niger's annual debt service obligations, freeing funds for expenditures on basic health care, primary education, HIV/AIDS prevention, rural infrastructure, and other programs geared at poverty reduction.

In December 2005, it was announced that Niger had received 100% multilateral debt relief from the IMF, which translates into the forgiveness of approximately $86 million USD in debts to the IMF, excluding the remaining assistance under HIPC. Nearly half of the government's budget is derived from foreign donor resources. Future growth may be sustained by exploitation of oil, gold, coal, and other mineral resources. Uranium prices have recovered somewhat in the last few years. A drought and locust infestation in 2005 led to food shortages for as many as 2.5 million Nigeriens.

Agriculture

The agricultural economy is based largely upon internal markets, subsistence agriculture, and the export of raw commodities: food stuffs and cattle to neighbours. Niger's agricultural and livestock sectors are the mainstay of all but 18% of the population. Fourteen percent of Niger's GDP is generated by livestock production (camels, goats, sheep and cattle), said to support 29% of the population. Thus 53% of the population is actively involved in crop production. The 15% of Niger's land that is arable is found mainly along its southern border with Nigeria.

In these areas, Pearl millet, sorghum, and cassava are the principal rain-fed subsistence crops. Irrigated rice for internal consumption is grown in parts of the Niger River valley in the west. While expensive, it has, since the devaluation of the CFA franc, sold for below the price of imported rice, encouraging additional production. Cowpeas and onions are grown for commercial export, as are small quantities of garlic, peppers, potatoes, and wheat. Oasis farming in small patches of the north of the country produces onions, dates, and some market vegetables for export.

But for the most part, rural residents engaged in crop tending are clustered in the south centre and south west of the nation, in those areas (the Sahel) which can expect to receive between 300mm to 600mm of rainfall annually. A small area in the southern tip of the nation, surrounding Gaya can expect to receive 700mm to 900mm or rainfall. Northern areas which support crops, such as the southern portions of the Air Massif and the Kaouar oasis rely upon oases and a slight increase in rainfall due to mountain effects. Large portions of the northwest and far east of the nation, while within the Sahara desert, see just enough seasonal rainfall to support semi-nomadic animal husbandry. The populations of these areas, mostly Tuareg, Wodaabe-Fula, and Toubou, travel south (a process called Transhumance) to pasture and sell animals in the dry season, north into the Sahara in the brief rainy season.

Rainfall varies and when insufficient, Niger has difficulty feeding its population and must rely on grain purchases and food aid to meet food requirements. Rains, as in much of the Sahel, have been marked by annual variability. This has been especially true in the 20th century, with the most severe drought on record beginning in the late 1960s and lasting, with one break, well into the 1980s. The long term effect of this, especially to pastoralist populations remains in the 21st century, with those communities which rely upon cattle, sheep, and camels husbandry losing entire herds more than once during this period. Recent rains remain variable. For instance, the rains in 2000 were not good, those in 2001 were plentiful and well distributed.

The Kandadji dam on the Niger River, whose construction started in August 2008, is expected to improve agricultural production in the Tillaberi Department by providing water for the irrigation of 6,000 hectares initially and of 45,000 hectares by 2034.

Exports

Uranium is Niger's largest export. Foreign exchange earnings from livestock, although difficult to quantify, are second. Actual exports far exceed official statistics, which often fail to detect large herds of animals informally crossing into Nigeria. Some hides and skins are exported, and some are transformed into handicrafts. Substantial deposits of phosphates, coal, iron, limestone, and gypsum also have been found in Niger.

Uranium

The persistent uranium price slump has brought lower revenues for Niger's uranium sector, although uranium still provides 72% of national export proceeds. The nation enjoyed substantial export earnings and rapid economic growth during the 1960s and 1970s after the opening of two large uranium mines near the northern town of Arlit. When the uranium-led boom ended in the early 1980s, however, the economy stagnated, and new investment since then has been limited. Niger's two uranium mines—SOMAIR's open pit mine and COMINAK's underground mine—are owned by a French-led consortium and operated by French interests. However, as of 2007, many licences have been given to other companies from countries such as Canada and Australia in order to exploit new deposits.

Gold

Exploitable deposits of gold are known to exist in Niger in the region between the Niger River and the border with Burkina Faso. On October 5, 2004, President Tandja announced the official opening of the Samira Hill Gold Mine in Tera Department and the first Nigerien gold ingot was presented to him. This marked a historical moment for Niger as the Samira Hill Gold Mine represents the first commercial gold production in the country.

Samira Hill is owned by a company called SML (Societe des Mines du Liptako) which is a joint venture between a Moroccan company, Societe Semafo, and a Canadian company, Etruscan Resources. Both companies own 80% (40%-40%) of SML and the Government of Niger 20%. The first year's production is predicted to be 135,000 troy ounces (4,200 kg; 9,260 lb avoirdupois) of gold at a cash value of USD 177 per ounce ($5.70/g). The mine reserves for the Samira Hill mine total 10,073,626 tons at an average grade of 2.21 grams per ton from which 618,000 troy ounces (19,200 kg; 42,400 lb) will be recovered over a 6 year mine life. SML believes to have a number of significant gold deposits within what is now recognized as the gold belt known as the "Samira Horizon", which is located between Gotheye and Ouallam.

Coal

The parastatal SONICHAR (Societe Nigerienne de Charbon) in Tchirozerine (north of Agadez) extracts coal from an open pit and fuels an electricity generating plant that supplies energy to the uranium mines. There are additional coal deposits to the south and west that are of a higher quality and may be exploitable.

Oil

Niger has oil potential. In 1992, the Djado permit was awarded to Hunt Oil Company, and in 2003 the Tenere permit was awarded to the China National Petroleum Company. An Exxon Mobil-Petronas joint venture was sold sole rights to the Agadem block, in the Diffa Region north of Lake Chad, but never went beyond exploration.

In June 2008, the government transferred the Agadem block rights to CNPC. Niger announced that in exchange for the USD$5 Billion investment, the Chinese company would build wells, 11 of which would open by 2012, a 20,000-barrel-per-day (3,200 m/d) refinery near Zinder and a pipeline out of the nation. The government estimates the area has reserves of 324 million barrels (51,500,000 m^3), and is seeking further oil in the Tenere Desert and near Bilma. Niger has said that it hopes to produce its first barrels of oil for sale by 2009.

Growth Rates

The economic competitiveness created by the January 1994 devaluation of the Communaute Financiere Africaine (CFA) franc contributed to an annual average economic growth of 3.5% throughout the mid-1990s. But the economy stagnated due to the sharp reduction in foreign aid in 1999 (which gradually resumed in 2000) and poor rains in 2000. Reflecting the importance of the agricultural sector, the return of good rains was the primary factor underlying economic growth of 5.1% in 2000, 3.1% in 2001, 6.0% in 2002, and 3.0% in 2003.

In recent years, the Government of Niger drafted revisions to the investment code (1997 and 2000), petroleum code (1992), and mining code (1993), all with attractive terms for investors. The present government actively seeks foreign private investment and considers it key to restoring economic growth and development. With the assistance of the United Nations Development Programme (UNDP), it has undertaken a concerted effort to revitalize the private sector.

Economic Restructuring and Debt

In January 2000, Niger's newly elected government inherited serious financial and economic problems including a virtually empty treasury, past-due salaries (11 months of arrears) and scholarship payments, increased debt, reduced revenue performance, and lower public investment. In December 2000, Niger qualified for enhanced debt relief under the International Monetary Fund program for Highly Indebted Poor Countries and concluded an agreement with the Fund on a Poverty Reduction and Growth Facility (PRGF).

In addition to changes in the budgetary process and public finances, the new government has pursued economic restructuring towards the IMF promoted privatization model. This has included the privatization of water distribution and telecommunications and the removal of price protections for petroleum products, allowing prices to be set by world market prices. Further privatizations of public enterprises are in the works. In its effort comply with the IMF's Poverty Reduction and Growth Facility plan, the government also is taking actions to reduce corruption and, as the result of a participatory process encompassing civil society, has devised a Poverty Reduction Strategy Plan that focuses on improving health, primary education, rural infrastructure, and judicial restructuring. A long planned privitisation of the Nigerien power company, NIGELEC, failed in 2001 and again in 2003 due to inaudibility to line up buyers.

SONITEL, the nation's telephone operator, hived of the post office and privatised in 2001, was renationalised in 2009.

Privatization and economic liberalization have however also been the subject of strong criticism. The UN Special Rapporteur on the Right to Food, for instance, has noted that privatization affects the poorest and most vulnerable members of Niger's society. Critics have argued that the obligations to creditor institutions and governments have locked Niger in to a process of trade liberalization that is harmful for small farmers and in particular, rural women.

Foreign Aid

The most important donors in Niger are France, the European Union, the World Bank, the IMF and other United Nations agencies (UNDP, UNICEF, FAO, WFP, and UNFPA). Other principal donors include the United States, Belgium, Germany, Switzerland, Canada, and Saudi Arabia. While USAID does not have an office in Niger, the United States is a major donor, contributing nearly $10 million each year to Niger's development.

The U.S. also is a major partner in policy coordination in such areas as food security and HIV/AIDS. The importance of external support for Niger's development is demonstrated by the fact that about 45% of the government's FY 2002 budget, including 80% of its capital budget, derives from donor resources.

In 2005 the UN drew attention to the increased need for foreign aid given severe problems with drought and locusts resulting in the 2005–06 Niger food crisis, endangering the lives around a million people.

Demographics

Over half the population of Niger belong to the Hausa, who also constitute the major ethnic group in northern Nigeria, and the Zarma-Songhai, who also are found in parts of Mali. Both groups, along with the Gourmantche, are sedentary farmers who live in the arable, southern tier of the country.

The remainder of Nigeriens are nomadic or semi-nomadic livestock-raising peoples—Fulani, Tuareg, Kanuri, Arabs, and Toubou—who make up about 20% of Niger's population. With rapidly growing populations and the consequent competition for meager natural resources, lifestyles of agriculturalists and livestock herders have come increasingly into conflict in Niger in recent years. A Nigerien study has found that more than 800,000 people are enslaved, almost 8% of the population.

Health

Niger's high infant mortality rate is comparable to levels recorded in neighbouring countries. However, the child mortality rate (deaths among children between the ages of 1 and 4) is exceptionally high (248 per 1,000) due to generally poor health conditions and inadequate nutrition for most of the country's children. According to the organization Save the Children, Niger has the world's highest infant mortality rate.

Nonetheless, Niger has the highest fertility rate in the world (7.2 births per woman); this means that nearly half (49%) of the Nigerien population is under age 15. There were 3 physicians and 22 nurses per 100,000 persons in 2004.

Education

Primary education in Niger is compulsory for six years. The primary school enrollment and attendance rates are low, particularly for girls. In 1997, the gross primary enrollment rate was 29.3 percent, and in 1996, the net primary enrollment rate was 24.5 percent. About 60 percent of children who finish primary schools are boys, as the majority of girls rarely attend school for more than a few years. Children are often forced to work rather than attend school, particularly during planting or harvest periods. In addition, nomadic children in the north of the country often do not have access to schools.

Culture and Religion

Nigerien culture is marked by variation, evidence of the cultural crossroads which French colonialism formed into a unified state from the beginning of the 20th century.

What is now Niger was created from four distinct cultural areas in the pre-colonial era: the Zarma dominated Niger River valley in the southwest; the northern periphery of Hausaland, made mostly of those states which had resisted the Sokoto Caliphate, and ranged along the long southern border with Nigeria; the Lake Chad basin and Kaouar in the far east, populated by Kanuri farmers and Toubou pastoralists who had once been part of the Kanem-Bornu Empire; and the Tuareg nomads of the Air Mountains and Saharan desert in the vast north.

Each of these communities, along with smaller ethnic groups like the pastoral Wodaabe Fula, brought their own cultural traditions to the new state of Niger. While successive post-independence governments have tried to forge a shared national culture, this has been slow forming, in part because the major Nigerien communities have their own cultural histories, and in part because Nigerien ethnic groups such as the Hausa, Tuareg and Kanuri are but part of larger ethnic communities which cross borders introduced under colonialism.

Until the 1990s, government and politics was inordinately dominated by Niamey and the Zarma people of the surrounding region. At the same time the plurality of the population, in the Hausa borderlands between Birni-N'Konni and Maine-Soroa, have often looked culturally more to Hausaland in Nigeria than Niamey. Between 1996 and 2003, primary school attendance was around 30%, including 36% of males and only 25% of females. Additional education occurs through madrassas.

Media

Niger began developing diverse media in the late 1990s. Prior to the Third Republic, Nigeriens only had access to tightly controlled state media. Now Niamey boasts scores of newspapers and magazines, some, like *Le Sahel*, are government operated, while many are critical of the government. Radio is the most important medium, as television sets are beyond the buying power of many of the rural poor, and illiteracy prevents print media from becoming a mass medium.

In addition to the national and regional radio services of the state broadcaster ORTN, there are four privately owned radio networks which total more than 100 stations. Three of them—the Anfani Group, Sarounia and Tenere—are urban based commercial format FM networks in the major towns. There is also a network of over 80 community radio stations spread across all seven regions of the country, governed by the Comite de Pilotage de Radios de Proximity (CPRP), a civil society organisation. The independent sector radio networks are collectively estimated by CPRP officials to cover some 7.6 million people, or about 73% of the population (2005).

Aside from Nigerien radio stations, the BBC's Hausa service is listened to on FM repeaters across wide parts of the country, particularly in the south, close to the border with Nigeria. Radio France Internationale also rebroadcasts in French through some of the commercial stations, via satellite. Tenere FM also runs a national independent television station of the same name.

Despite relative freedom at the national level, Nigerien journalists say they are often pressured by local authorities. The state ORTN network depends financially on the government, partly through an addition to electricity bills and partly through direct subsidy. The sector is governed by the Conseil Superieur de Communications, established as an independent body in the early 1990s, since 2007 headed by Daouda Diallo. International human rights groups have criticised government since at least 1996 as using regulation and police to punish criticism of the state.

7

Nigeria

Nigeria, officially the Federal Republic of Nigeria, is a federal constitutional republic comprising thirty-six states and its Federal Capital Territory, Abuja. The country is located in West Africa and shares land borders with the Republic of Benin in the west, Chad and Cameroon in the east, and Niger in the north. Its coast in the south lies on the Gulf of Guinea on the Atlantic Ocean. The three largest and most influential ethnic groups in Nigeria are the Hausa, Igbo and Yoruba. In terms of religion Nigeria is roughly split half and half between Muslims and Christians with a very small minority who practice traditional religions.

The people of Nigeria have an extensive history. Archaeological evidence shows that human habitation of the area dates back to at least 9000 BC. The area around the Benue and Cross River is thought to be the original homeland of the Bantu migrants who spread across most of central and southern Africa in waves between the 1st millennium BC and the 2nd millennium.

The name *Nigeria* was taken from the Niger River running through the country. This name was coined by Flora Shaw, the future wife of Baron Lugard, a British colonial administrator, in the late 19th century.

Nigeria is the most populous country in Africa, the eighth most populous country in the world, and the most populous country in the world in which the majority of the population is black. It is listed among the "Next Eleven" economies, and is a member of the Commonwealth of Nations. The economy of Nigeria is one of the fastest growing in the world, with the International Monetary Fund projecting a growth of 9% in 2008 and 8.3% in 2009. It is the third largest economy in Africa, and is a regional power that is also the hegemon in West Africa.

History

Early History

The Nok people of central Nigeria produced the earliest terracotta sculptures ever to be found in the country. A Nok sculpture resident at the Minneapolis Institute of Arts, portrays a sitting dignitary wearing a "Shepherds Crook" on the right arm, and a "hinged flail" on the left. These are symbols of authority associated with ancient Egyptian pharaohs, and the god Osiris, and suggests that an ancient Egyptian style of social structure, and perhaps religion, existed in the area of modern Nigeria during the late Pharonic period.

In the northern part of the country, Kano and Katsina had recorded history dateing back to around 999. Hausa kingdoms and the Kanem-Bornu Empire prospered as trade posts between North and West Africa. At the beginning of the 19th century under Usman dan Fodio the Fulani leaded the centralized Fulani Empire which continued until 1903 when the Fulani population and land were divided into various European colonies. Between 1750 and 1900, between one to two-thirds of the entire population of the Fulani jihad states consisted of slaves.

The Yoruba kingdoms of If[1] and Oyo in the southwestern block of Nigeria became prominent around 700—900 and 1400 respectively. However, Yoruba mythology states that Ile-Ife is the source of the human race and that it predates any other civilization. If[1] also produced terra cotta and bronze figures and Oyo once extended from western Nigeria to Togo. Arguably the most powerful and prominent kingdom in the whole of Nigeria's history was also located in southwestern Nigeria, the Kingdom of Benin. Benin's power lasted between the 15th and 19th century. Their dominance reached as far as the city of Eko (a Bini name later changed to Lagos by the Portuguese) and further.

In southeastern Nigeria the Kingdom of Nri of the Igbo people flourished from the controversial date of around the 10th century until 1911, making it the oldest kingdom in Nigeria. The Nri Kingdom was ruled by the Eze Nri. The city of Nri is considered to be the foundation of Igbo culture. Nri and Aguleri, where the Igbo creation myth originates, are in the territory of the Umeuri clan, who trace their lineages back to the patriarchal king-figure, Eri.

Colonial Era

Portuguese explorers were the first Europeans to begin trade in Nigeria in the port they named Lagos and in Calabar. The Europeans traded with the ethnicities of the coast and also negotiated a trade in slaves, to the detriment and profit of many Nigerian ethnicities. Consequently many of the citizens of the former slave nations of the British Empire are descended from a Nigerian ethnic group. Britain abolished its slave trade in 1807 and, following the Napoleonic Wars, established the West Africa Squadron in an attempt to halt the international traffic in slaves.

In 1885 British claims to a West African sphere of influence received international recognition and in the following year the Royal Niger Company was chartered under the leadership of Sir George Taubman Goldie. In 1900 the company's territory came under the control of the British government, which moved to consolidate its hold over the area of modern Nigeria. On January 1, 1901 Nigeria became a British protectorate, part of the British Empire, the foremost world power at the time. Many wars against subjugation had been fought by the states of what later became Nigeria against the British Empire in the late nineteenth and early twentieth century. Notably of those were the British Conquest of Benin in 1897 and the Anglo-Aro War from 1901—1902. The restraint or complete destruction of these states opened up the Niger area to British rule.

In 1914, the Niger area was formally united as the *Colony and Protectorate of Nigeria*. Administratively, Nigeria remained divided into the northern and southern provinces and Lagos colony. Western education and the development of a modern economy proceeded more rapidly in the south than in the north, with consequences felt in Nigeria's political life ever since.

Slavery was not finally outlawed in northern Nigeria until 1936.

Following World War II, in response to the growth of Nigerian nationalism and demands for independence, successive constitutions legislated by the British Government moved Nigeria toward self-government on a representative and increasingly federal basis. By the middle of the 20th century, the great wave for independence was sweeping across Africa.

Post-independence

On October 1, 1960, Nigeria gained its independence from the United Kingdom. The new republic incorporated a number of people with aspirations of their own sovereign nations. Newly independent, Nigeria's government was a coalition of conservative parties: the Nigerian People's Congress (NPC), a party dominated by Northerners and those of the Islamic faith, and the Igbo and Christian dominated National Council of Nigeria and the Cameroons (NCNC) led by Nnamdi Azikiwe, who became Nigeria's maiden Governor-General in 1960. Forming the opposition was the comparatively liberal Action Group (AG), which was largely dominated by the Yoruba and led by Obafemi Awolowo. The cultural and political differences between Nigeria's dominant ethnicities, the Hausa ('Northerners'), Igbo ('Easterners') and Yoruba ('Westerners'), were sharp.

An imbalance was created in the polity by the result of the 1961 plebiscite. Southern Cameroon opted to join the Republic of Cameroon while northern Cameroon chose to remain in Nigeria. The northern part of the country was now far larger than the southern part. The nation parted with its British legacy in 1963 by declaring itself a Federal Republic, with Azikiwe as its first president. When elections came about in 1965, the AG was outmanoeuvred for control of Nigeria's Western Region by the Nigerian National Democratic Party, an amalgamation of conservative Yoruba elements backed heavily by the Federal Government amid dubious electoral circumstances.

Nigerian-Biafran War

The disequilibrium and perceived corruption of the electoral and political process led in 1966 to several back-to-back military coups. The first was in January and led by a collection of young leftists under Major Emmanuel Ifeajuna and Chukwuma Kaduna Nzeogwu. It was partially successful; the coup plotters murdered the Prime Minister, Sir Abubakar Tafawa Balewa, the Premier of the Northern Region, Sir Ahmadu Bello, and the Premier of the Western Region, Sir Ladoke Akintola.

Despite this, the coupists could not set up a central government because of logistic reasons. Sir Nwafor Orizu, the acting President was then pressured to hand over government to the Nigeria Army, under the Command of General JTU Aguyi-Ironsi.

The coup was counter-acted by another successful plot, supported primarily by Northern military officers and Northerners who favoured the NPC, it was engineered by Northern officers, which allowed Lt Colonel Yakubu Gowon to become head of state. This sequence of events led to an increase in ethnic tension and violence. The Northern coup, which was mostly motivated by ethnic and religious reasons was a bloodbath of both military officers and civilians, especially those of Igbo extraction.

The violence against the Igbo increased their desire for autonomy and protection from the military's wrath. By May 1967, the Eastern Region had declared itself an independent state called the Republic of Biafra under the leadership of Lt Colonel Emeka Ojukwu in line with the wishes of the people.

The Nigerian Civil War began as the Nigerian (Western and Northern) side attacked Biafra (South-eastern) on July 6, 1967 at Garkem signalling the beginning of the 30 month war that ended in January 1970. More than one million people died in the three-year civil war.

Following the war, Nigeria became to an extent even more mired in ethnic strife, as the defeated southeast and indeed southern Nigeria was now conquered territory for the federal military regime, which changed heads of state twice, as army officers staged a bloodless coup against Gowon and enthroned Murtala Mohammed; Olusegun Obansanjo succeeded the former after an assassination.

Military Era

During the oil boom of the 1970s, Nigeria joined OPEC and billions of dollars generated by production in the oil-rich Niger Delta flowed into the coffers of the Nigerian state. However, increasing corruption and graft at all levels of government squandered most of these earnings. The northern military clique benefited immensely from the oil boom to the detriment of the Nigerian people and economy. As oil revenues fuelled the rise of federal subventions to states and precariously to individuals, the Federal Government soon became the centre of political struggle and the centre became the threshold of power in the country. As oil production and revenue rose, the Nigerian government created a dangerous situation as it became increasingly dependent on oil revenues and the international commodity markets for budgetary and economic concerns eschewing economic stability. That spelled doom to federalism in Nigeria.

Beginning in 1979, Nigerians participated in a brief return to democracy when Obasanjo transferred power to the civilian regime of Shehu Shagari. The Shagari government was viewed as corrupt and incompetent by virtually all sectors of Nigerian society, so when the regime was overthrown by the military coup of Mohammadu Buhari shortly after the regime's fraudulent re-election in 1984, it was generally viewed as a positive development by most of the population. Buhari promised major reforms but his government fared little better than its predecessor, and his regime was overthrown by yet another military coup in 1985.

The new head of state, Ibrahim Babangida, promptly declared himself President and Commander in chief of the Armed Forces and the ruling Supreme Military Council and also set 1990 as the official deadline for a return to democratic governance. Babangida's tenure was marked by a flurry of political activity: he instituted the International Monetary Fund's Structural Adjustment Program (SAP) to aid in the repayment of the country's crushing international debt, which most federal revenue was dedicated to servicing. He also inflamed religious tensions in the nation and particularly the south by enrolling Nigeria in the Organization of the Islamic

Conference. After Babangida survived an abortive coup, he pushed back the promised return to democracy to 1992. When free and fair elections were finally held on the 12th of June, 1993, Babangida declared that the results showing a presidential victory for Moshood Kashimawo Olawale Abiola null and void, sparking mass civilian violence in protest which effectively shut down the country for weeks and forced Babangida to keep his shaky promise to relinquish office to a civilian run government. Babangida's regime is adjudged to be at the apogee of corruption in the history of the nation as it was during his time that corruption became officially diluted in Nigeria.

Babangida's caretaker regime headed by Ernest Shonekan survived only until late 1993 when General Sani Abacha took power in another military coup. Abacha proved to be perhaps Nigeria's most brutal ruler and employed violence on a wide scale to suppress the continuing pandemic of civilian unrest. Money had been found in various western European countries banks traced to him. He avoided coup plots by bribing army generals. Several hundred millions dollars in accounts traced to him were unearthed in 1999. The regime would come to an end in 1998 when the dictator was found dead amid dubious circumstances. Abacha's death yielded an opportunity for return to civilian rule.

Recent History

Nigeria re-achieved democracy in 1999 when it elected Olusegun Obasanjo, the former military head of state, as the new President ending almost thirty three-years of military rule (from 1966 until 1999) excluding the short-lived second republic (between 1979 and 1983) by military dictators who seized power in coups d'etat and counter-coups during the Nigerian military juntas of 1966-1979 and 1983-1998.

Although the elections which brought Obasanjo to power in 1999 and again in 2003 were condemned as unfree and unfair, Nigeria has shown marked improvements in attempts to tackle government corruption and to hasten development. Whiie Obasanjo showed willingness to fight corruption, he was accused by others of the same. Umaru Yar Adua, of the People's Democratic Party, came into power in the general election of

2007 – an election that was witnessed and condemned by the international community as being massively flawed.

Ethnic violence over the oil producing Niger Delta region and inadequate infrastructures are some of the current issues in the country.

Yar' Adua died on 5 May 2010. Dr. Goodluck Ebele Jonathan was sworn in as Yar' Adua's replacement on 6 May 2010, becoming Nigeria's 14th Head of State. He will serve as President until the next election. Upon taking office, Jonathan cited anti-corruption, power and electoral reform as likely focuses of his administration. He stated that he came to office under "very sad and unusual circumstances."

On 18 May 2010, the National Assembly approved President Goodluck Jonathan's nomination of former Kaduna State governor, Namadi Sambo, an architect, for the position of Vice President of the Federal Republic of Nigeria.

Government and Politics

Nigeria is a Federal Republic modelled after the United States, with executive power exercised by the president and with overtones of the Westminster System model in the composition and management of the upper and lower houses of the bicameral legislature. The current president of Nigeria is Goodluck Jonathan, who succeeded to the office in 2010. The president presides as both Head of State and head of the national executive and is elected by popular vote to a maximum of two four-year terms.

The president's power is checked by a Senate and a House of Representatives, which are combined in a bicameral body called the National Assembly. The Senate is a 109-seat body with three members from each state and one from the capital region of Abuja; members are elected by popular vote to four-year terms. The House contains 360 seats and the number of seats per state is determined by population.

Ethnocentrism, tribalism, religious persecution, and prebendalism have played a visible role in Nigerian politics both prior and subsequent to independence in 1960. Kin-selective altruism has made its way into Nigerian politics and has spurned various attempts by tribalists to concentrate Federal power to a particular region of their interests. Nationalism has also led to active secessionist movements such as MASSOB, Nationalist movements such as Oodua Peoples Congress, Movement for the Emancipation of the Niger Delta and a civil war. Nigeria's three largest ethnic groups (Hausa, Igbo and Yoruba) have maintained historical preeminence in Nigerian politics; competition amongst these three groups has fuelled corruption and graft.

Because of the above issues, Nigeria's current political parties are pan-national and irreligious in character (though this does not preclude the continuing preeminence of the dominant ethnicities). The major political parties at present include the ruling People's Democratic Party of Nigeria which maintains 223 seats in the House and 76 in the Senate (61.9% and 69.7% respectively); the opposition All Nigeria People's Party under the leadership of Muhammadu Buhari has 96 House seats and 27 in the Senate (26.6% and 24.7%). There are also about twenty other minor opposition parties registered. The immediate past president, Olusegun Obasanjo, acknowledged fraud and other electoral "lapses" but said the result reflected opinion polls. In a national television address he added that if Nigerians did not like the victory of his handpicked successor they would have an opportunity to vote again in four years.

Like in many other African societies, prebendalism and extremely excessive corruption continue to constitute major challenges to Nigeria, as vote rigging and other means of coercion are practised by all major parties in order to remain competitive. In 1983, it was adjudged by the policy institute at Kuru that only the 1959 and 1979 elections witnessed minimal rigging.

Law

There are four distinct systems of law in Nigeria:

- English law which is derived from its colonial past with Britain;
- Common law, a development of its post colonial independence;
- Customary law which is derived from indigenous traditional norms and practice, including the dispute

resolution meetings of pre-colonial Yorubaland secret societies and the Ekpe and Okonko of Igboland and Ibibioland;

- Sharia law, used only in the predominantly Muslim north of the country. It is an Islamic legal system which had been used long before the colonial administration in Nigeria but recently politicised and spearheaded in Zamfara in late 1999 and eleven other states followed suit. These states are Kano, Katsina, Niger, Bauchi, Borno, Kaduna, Gombe, Sokoto, Jigawa, Yobe, and Kebbi.

The country has a judicial branch, the highest court of which is the Supreme Court of Nigeria.

Foreign Relations

Upon gaining independence in 1960, Nigeria made the liberation and restoration of the dignity of Africa the centrepiece of its foreign policy and played a leading role in the fight against the apartheid regime in South Africa. One notable exception to the African focus of Nigeria's foreign policy was the close relationship the country enjoyed with Israel throughout the 1960s, with the latter country sponsoring and overseeing the construction of Nigeria's parliament buildings.

Nigeria's foreign policy was soon tested in the 1970s after the country emerged united from its own civil war and quickly committed itself to the liberation struggles going on in the Southern Africa sub-region.

Though Nigeria never sent an expeditionary force in that struggle, it offered more than rhetoric to the African National Congress (ANC) by taking a committed tough line with regard to the racist regime and their incursions in southern Africa, in addition to expediting large sums to aid anti-colonial struggles. Nigeria was also a founding member of the Organisation for African Unity (now the African Union), and has tremendous influence in West Africa and Africa on the whole. Nigeria has additionally founded regional cooperative efforts in West Africa, functioning as standard-bearer for ECOWAS and ECOMOG, economic and military organizations respectively.

With this African-centred stance, Nigeria readily sent troops to the Congo at the behest of the United Nations shortly after independence (and has maintained membership since that time); Nigeria also supported several Pan African and pro-self government causes in the 1970s, including garnering support for Angola's MPLA, SWAPO in Namibia, and aiding anti-colonial struggles in Mozambique, and Zimbabwe (then Rhodesia) military and economically.

Nigeria retains membership in the Non-Aligned Movement, and in late November 2006 organized an Africa-South America Summit in Abuja to promote what some attends termed "South-South" linkages on a variety of fronts. Nigeria is also a member of the International Criminal Court, and the Commonwealth of Nations, from which it was temporarily expelled in 1995 under the Abacha regime.

Nigeria has remained a key player in the international oil industry since the 1970s, and maintains membership in Organization of the Petroleum Exporting Countries OPEC which it joined in July, 1971. Its status as a major petroleum producer figures prominently in its sometimes vicissitudinous international relations with both developed countries, notably the United States and more recently China and developing countries, notably Ghana, Jamaica and Kenya.

Millions of Nigerians have emigrated at times of economic hardship to Europe, North America and Australia among others. It is estimated that over a million Nigerians have emigrated to the United States and constitute the Nigerian American populace. Of such Diasporic communities include the "Egbe Omo Yoruba" society.

Military

The Nigerian Military are charged with protecting The Federal Republic of Nigeria, promoting Nigeria's global security interests, and supporting peacekeeping efforts especially in West Africa.

The Nigerian Military consist of an Army, a Navy and an Air Force. The military in Nigeria have played a major role in the country's history since independence. Various juntas have seized control of the country and ruled it through most of its history. Its last period of rule ended in 1999 following the sudden death of former dictator Sani Abacha in 1998, with his successor, Abdulsalam Abubakar, handing over power to the democratically elected government of Olusegun Obasanjo in 1999.

Taking advantage of its role as Africa's most populated country, Nigeria has repositioned its military as an African peacekeeping force. Since 1995, the Nigerian military through ECOMOG mandates have been deployed as peacekeepers in Liberia (1997), Ivory Coast (1997–1999), Sierra Leone 1997–1999, and presently in Sudan's Darfur region under an African Union mandate.

Geography

Nigeria is located in western Africa on the Gulf of Guinea and has a total area of 923,768 km^2 (356,669 sq mi), making it the world's 32nd-largest country (after Tanzania). It is comparable in size to Venezuela, and is about twice the size of California. It shares a 4,047 kilometres (2,515 mi) border with Benin (773 km), Niger (1497 km), Chad (87 km), Cameroon (1690 km), and has a coastline of at least 853 km. The highest point in Nigeria is Chappal Waddi at 2,419 m (7,936 ft). The main rivers are the Niger and the Benue which converge and empty into the Niger Delta, one of the world's largest river deltas and the location of a large area of Central African Mangroves.

Nigeria is also an important centre for biodiversity. It is widely believed that the areas surrounding Calabar, Cross River State, contain the world's largest diversity of butterflies. The drill monkey is only found in the wild in Southeast Nigeria and neighbouring Cameroon.

Nigeria has a varied landscape. The far south is defined by its tropical rainforest climate, where annual rainfall is 60 to 80 inches (1,524 to 2,032 mm) a year. In the southeast stand the Obudu plateau. Coastal plains are found in both the southwest and the southeast. This forest zone's most southerly portion is defined as salt water swamp, also known as a mangrove swamp because of the large amount of mangroves in the area. North of this is fresh water swamp, containing different vegetation from the salt water swamp, and north of that is rain forest.

Nigeria's most expansive topographical region is that of the valleys of the Niger and Benue River valleys (which merge into each other and form a "y" shape). To the southwest of the Niger there is "rugged" highland, and to the southeast of the Benue are hills and mountains which forms the Mambilla plateau, the highest Plateau in Nigeria. This plateau extends to the border with Cameroon, this montane land is part of the Bamenda Highlands in Cameroon. The area near the border with Cameroon close to the coast is rich rainforest and part of the Cross-Sanaga-Bioko coastal forests ecoregion, an important centre for biodiversity including the drill monkey which is only found in the wild in this area and across the border in Cameroon. It is widely believed that the areas surrounding Calabar, Cross River State, also in this forest, contain the world's largest diversity of butterflies. The area of southern Nigeria between the Niger and the Cross Rivers has seen its forest more or less disappear to be replaced by grassland.

Everything in between the far south and the far north, is savannah (insignificant tree cover, with grasses and flowers located between trees), and rainfall is between 20 and 60 inches (508 and 1,524 mm) per year. The savannah zone's three categories are Guinean forest-savanna mosaic, plains of tall grass which are interrupted by trees and the most common across the country: Sudan savannah, similar but with "shorter grasses and shorter trees; and Sahel savannah, comprised patches of grass and sand, found in the northeast. To the north is the Sahel with its almost desert-like climate, where rain is less than 20 inches (508 mm) per year and the Sahara Desert is encroaching. In the dry north-east corner of the country lies Lake Chad, which Nigeria shares with Niger, Chad and Cameroon.

Environmental Issues

Nigeria's Delta region, home of the large oil industry, experiences serious oil spills and other environmental problems.

Waste management including sewage treatment, the linked processes of deforestation and soil degradation, and climate change or global warming are the major environmental problems in Nigeria.

Waste management presents problems in a mega city like Lagos and other major Nigerian cities which are linked with economic development, population growth and the inability of municipal councils to manage the resulting rise in industrial and domestic waste.

Haphazard industrial planning, increased urbanization, poverty and lack of competence of the municipal government are seen as the major reasons for high levels of waste pollution in major Nigerian cities. Some of the 'solutions' have been disastrous to the environment, resulting in untreated waste being dumped in places where it can pollute waterways and groundwater.

In terms of global warming, Africans contribute only about one metric ton of carbon dioxide per person per year. It is perceived by many climate change experts that food production and security in the northern sahel region of the country will suffer as semi-arid areas will have more dry periods in the future.

Subdivisions

Nigeria is divided into thirty-six states and one Federal Capital Territory, which are further sub-divided into 774 Local Government Areas (LGAs).

The plethora of states, of which there were only three at independence, reflect the country's tumultuous history and the difficulties of managing such a heterogeneous national entity at all levels of government.

Nigeria has six cities with a population of over 1 million people (from largest to smallest: Lagos, Kano, Ibadan, Kaduna, Port Harcourt, and Benin City). Lagos is the largest city in sub-Saharan Africa, with a population of over 10 million in its urban area alone. Population of Nigeria's cities over a million include Lagos (7,937,932), Kano (3,848,885), Ibadan (3,078,400), Kaduna (1,652,844), Port Harcourt (1,320,214), Benin City (1,051,600), Maiduguri (1,044,497) and Zaria (1,018,827), however, these figures are regularly disputed in Nigeria.

Economy

Nigeria is classified as an emerging market, and has already reached middle income status according to the Worldbank, with its abundant supply of natural resources, well-developed financial, legal, communications, transport sectors and stock exchange (the Nigerian Stock Exchange), which is the second largest in Africa. Nigeria is ranked 37th in the world in terms of GDP (PPP) as of 2007. Nigeria is the United States' largest trading partner in sub-Saharan Africa and supplies a fifth of its oil (11% of oil imports). It has the seventh-largest trade surplus with the U.S. of any country worldwide. Nigeria is currently the 50th-largest export market for U.S. goods and the 14th-largest exporter of goods to the U.S. The United States is the country's largest foreign investor.

Previously, economic development had been hindered by years of military rule, corruption, and mismanagement, the restoration of democracy and subsequent economic reforms have successfully put Nigeria back on track towards achieving its full economic potential. It is now one of the largest economies in Africa, and the largest economy in the West Africa Region.

According to the Economist Intelligence Unit and the World Bank, Nigerian GDP at purchasing power parity has nearly doubled from $170.7 billion in 2005 to 292.6 billion in 2007. The GDP per head has jumped from $692 per person in 2006 to $1,754 per person in 2007.

During the oil boom of the 1970s, Nigeria accumulated a significant foreign debt to finance major infrastructural investments. With the fall of oil prices during the 1980s oil glut Nigeria struggled to keep up with its loan payments and eventually defaulted on its principal debt repayments, limiting repayment to the interest portion of the loans. Arrears and penalty interest accumulated on the unpaid principal which increased the size of the debt.

However, after negotiations by the Nigeria authorities, in October 2005 Nigeria and its Paris Club creditors reached an agreement in which Nigeria repurchased its debt at a discount of approximately 60%. Nigeria used part of its oil profits to pay the residual 40%, freeing up at least $1.15 billion annually for poverty reduction programmes. Nigeria made history in April 2006 by becoming the first African Country to completely pay off its debt (estimated $30 billion) owed to the Paris Club.

Key Sectors

Nigeria is the 12th largest producer of petroleum in the world and the 8th largest exporter, and has the 10th largest proven reserves. (The country joined OPEC in 1971). Petroleum plays a large role in the Nigerian economy, accounting for 40% of GDP and 80% of Government earnings. However, agitation for better resource control in the Niger Delta, its main oil producing region, has led to disruptions in oil production and currently prevents the country from exporting at 100% capacity.

Nigeria has one of the fastest growing telecommunications markets in the world, major emerging market operators (like MTN, Etisalat, Zain and Globacom) basing their largest and most profitable centres in the country. The government has recently begun expanding this infrastructure to space based communications. Nigeria has a space satellite which is monitored at the Nigerian National Space Research and Development Agency Headquarters in Abuja.

The country has a highly developed financial services sector, with a mix of local and international banks, asset management companies, brokerage houses, insurance companies and brokers, private equity funds and investment banks.

Nigeria also has a wide array of underexploited mineral resources which include natural gas, coal, bauxite, tantalite, gold, tin, iron ore, limestone, niobium, lead and zinc. Despite huge deposits of these natural resources, the mining industry in Nigeria is still in it infancy.

Agriculture used to be the principal foreign exchange earner of Nigeria. At one time, Nigeria was the world's largest exporter of groundnuts, cocoa, and palm oil and a significant producer of coconuts, citrus fruits, maize, pearl millet, cassava, yams and sugar cane. About 60% of Nigerians work in the agricultural sector, and Nigeria has vast areas of underutilized arable land.

It also has a manufacturing industry which includes leather and textiles (centred Kano, Abeokuta, Onitsha, and Lagos), car manufacturing (for the French car manufacturer Peugeot as well as for the English truck manufacturer Bedford, now a subsidiary of General Motors), t-shirts, plastics and processed food.

The country has recently made considerable amount of revenue from home made Nigerian Movies which are sold locally and Internationally. These movies are popular in other African countries and among African immigrants in Europe.

Demographics

Nigeria is the most populous country in Africa but exactly how populous is a subject of speculation. The United Nations estimates that the population in 2009 was at 154,729,000, distributed as 51.7% rural and 48.3% urban, and with a population density of 167.5 people per square kilometre. National census results in the past few decades have been disputed. The results of the most recent census were released in December 2006 and gave a population of 140,003,542. The only breakdown available was by gender: males numbered 71,709,859, females numbered 68,293,08.

According to the United Nations, Nigeria has been undergoing explosive population growth and one of the highest growth and fertility rates in the world. By their projections, Nigeria will be one of the countries in the world that will account for most of the world's total population increase by 2050.

According to current data, one out of every four Africans is Nigerian. Presently, Nigeria is the eighth most populous country in the world, and even conservative estimates conclude that more than 20% of the world's black population lives in Nigeria. 2006 estimates claim 42.3% of the population is between 0–14 years of age, while 54.6% is between 15–65; the birth rate is significantly higher than the death rate, at 40.4 and 16.9 per 1000 people respectively.

Health, health care, and general living conditions in Nigeria are poor. Life expectancy is 47 years (average male/female) and just over half the population has access to potable water and appropriate sanitation; the percentage is of children under five has gone up rather than down between 1990 and 2003 and infant mortality is 97.1 deaths per 1000 live births. HIV/AIDS rate in Nigeria is much lower compared to the other African nations such as Kenya or South Africa whose prevalence (percentage) rates are in the double digits. In 2003, the HIV prevalence rate among 20 to 29 year-olds was 5.6%. About Nigeria, like many developing countries, suffers from a polio crisis as well as periodic outbreaks of cholera, malaria, and sleeping sickness. As of 2004, there has been a vaccination drive, spearheaded by the W.H.O., to combat polio and malaria that has been met with controversy in some regions.

Education is also in a state of neglect. After the 1970s oil boom, tertiary education was improved so that it would reach every subregion of Nigeria. Education is provided free by the government, but the attendance rate for secondary education is only 29% (32% for males, 27% for females). The education system has been described as "dysfunctional" largely because of decaying institutional infrastructure. 68% of the population is literate, and the rate for men (75.7%) is higher than that for women (60.6%).

Nigeria's largest city is Lagos. Lagos has grown from 300,000 in 1950 to an estimated 15 million today, and the Nigerian government estimates that city will have expanded to 25 million residents by 2015.

Ethno-linguistic Groups

Nigeria has more than 250 ethnic groups, with varying languages and customs, creating a country of rich ethnic diversity. The largest ethnic groups are the Fulani/Hausa, Yoruba, Igbo, accounting for 68% of population, (Quote Source) while the Edo, Ijaw, Kanuri, Ibibio, Ebira Nupe and Tiv comprise 27%; (Quote Source?) other minorities make up the remaining 7 percent. The middle belt of Nigeria is known for its diversity of ethnic groups, including the Pyem, Goemai, and Kofyar. The official population count of each of Nigeria's ethnicities has always remained controversial and disputed as members of different ethnic groups believe the census is rigged to give a particular group (usually believed to be northern groups) numerical superiority.

There are small minorities of British, American, East Indian, Chinese (est. 50,000), white Zimbabwean, Japanese, Greek,

Syrian and Lebanese immigrants in Nigeria. Immigrants also include those from other West African or East African nations. These minorities mostly reside in major cities such as Lagos and Abuja, or in the Niger Delta as employees for the major oil companies. A number of Cubans settled in Nigeria as political refugees following the Cuban Revolution.

In the middle of the nineteenth century, a number of ex-slaves of Afro-Cuban and Afro-Brazilian descent and emigrants from Sierra Leone established communities in Lagos and other regions of Nigeria. Many ex-slaves came to Nigeria following the emancipation of slaves in the Americas. Many of the immigrants, sometimes called Saros (immigrants from Sierra Leone) and Amaro (ex-slaves from Brazil) later became prominent merchants and missionaries in these cities.

Language

The number of languages currently estimated and catalogued in Nigeria is 521. This number includes 510 living languages, two second languages without native speakers and nine extinct languages. In some areas of Nigeria, ethnic groups speak more than one language. The official language of Nigeria, English, was chosen to facilitate the cultural and linguistic unity of the country. The choice of English as the official language was partially related to the fact that a part of the Nigerian population spoke English as a result of British colonization that ended in 1960.

The major languages spoken in Nigeria represent three major families of African languages – the majority are Niger-Congo languages, such as Yoruba, Igbo, the Hausa language is Afro-Asiatic; and Kanuri, spoken in the northeast, primarily Borno State, is a member of the Nilo-Saharan family. Even though most ethnic groups prefer to communicate in their own languages, English, being the official language, is widely used for education, business transactions and for official purposes. English as a first language, however, remains an exclusive preserve of a small minority of the country's urban elite, and it is not spoken at all in some rural areas. With the majority of Nigeria's populace in the rural areas, the major languages of communication in the country remain indigenous languages.

Some of the largest of these, notably Yoruba and Igbo, have derived standardized languages from a number of different dialects and are widely spoken by those ethnic groups. Nigerian Pidgin English, often known simply as 'Pidgin' or 'Broken' (Broken English), is also a popular lingua franca, though with varying regional influences on dialect and slang. The pidgin English or Nigerian English is widely spoken within the Niger Delta Regions, predominately in Warri, Sapele, Port Harcourt, Agenebode, Ewu, and Benin City.

Culture

Literature

Nigerian citizens have authored many influential works of post-colonial literature in the English language. Nigeria's best-known writers are Wole Soyinka, the first African Nobel Laureate in Literature, and Chinua Achebe, best known for the novel, *Things Fall Apart* and his controversial critique of Joseph Conrad. Other Nigerian writers and poets who are well known internationally include John Pepper Clark, Ben Okri, Cyprian Ekwensi, Buchi Emecheta, Helon Habila, Chimamanda Ngozi Adichie, Ken Saro Wiwa, who was executed in 1995 by the military regime, and Saintmoses Eromosele a poet, playwright and social critic who wrote his first book, a novela, at the age of 16 and while still a secondary school student. Nigeria has the second largest newspaper market in Africa (after Egypt) with an estimated circulation of several million copies daily in 2003.

Music and Film

Nigeria has a role in the development of West African highlife, afrobeat, and palm-wine music, which fuses native rhythms with techniques imported from the Congo, Brazil, Cuba, and elsewhere.

Many late 20th century musicians such as Fela Kuti have famously fused cultural elements of various indigenous music with American Jazz and Soul to form Afrobeat. JuJu music which is percussion music fused with traditional music from the Yoruba nation and made famous by King Sunny Ade, is also from Nigeria. There is also fuji music, a Yoruba percussion style, created and popularized by Mr. Fuji, Alhaji Sikiru Ayinde Barrister. The is also the Afan Music invented and popularized by the Ewuborn poet and musician Umuobuarie Igberaese.

There is a budding hip hop movement in Nigeria. Kennis Music, the self-proclaimed number-one record label in Africa, and one of Nigeria's biggest record labels, has a roster almost entirely dominated by hip hop artists.

Some famous musicians that come from Nigeria are Fela Kuti, Adewale Ayuba, Ezebuiro Obinna, Alhaji Sikiru Ayinde Barrister, King Sunny Ade, Ebenezer Obey, Umobuarie Igberaese, Femi Kuti, Lagbaja, Dr. Alban, Sade Adu, Wasiu Alabi, Bola Abimbola and Tuface Idibia.

In November 2008, Nigeria's music scene (and that of Africa) received international attention when MTV hosted the continent's first African music awards show in Abuja.

The Nigerian video-film industry is known as Nollywood, which is now the second-largest producer of movies in the world. Many of the film studios are based in Lagos and Enugu, and the industry is now a very lucrative income for these cities.

Religion

Nigeria is home to a variety of religions which tend to vary regionally. This situation accentuates regional and ethnic distinctions and has often been seen as a source of sectarian conflict amongst the population. The largest religions of Nigeria are Islam and Christianity, including many followers of indigenous religions. Even though, Nigeria is apparently divided equally between Islam and Christianity between north and south, it is evident that across Nigeria there is widespread belief, albeit suppressed for political reasons, in traditional religious practices. Based on a 2003 survey, 50.5% were Muslim, 48.2% were Christian (15% Protestant, 13.7% Catholic, and 19.6% other Christian), and followers of other religions were 1.4%. The core north is largely Muslim, there are large numbers of both Muslims and Christians in the Middle Belt, including the Federal Capital Territory. In the southwest, Christians and Muslims reside unequally, southern regions are predominantly Christian with widespread traditional beliefs, in the east, Catholics, Anglicans, and Methodists are the majority with few traditional beliefs, while the Niger Delta region is mainly Christian.

The majority of Nigerian Muslims are Sunni, but a significant Shia and Sufi minority exists and a small minority of Ahmadiyya. Some northern states have incorporated Sharia law into their previously secular legal systems, which has brought about some controversy. Kano State has sought to incorporate Sharia law into its constitution.

Christian Nigerians are about evenly split between Roman Catholicism and Protestantism. Leading Protestant churches are the Church of Nigeria, of the Anglican communion, Assemblies of God Church, Nigeria, Redeemed Christian Church of God and the Nigerian Baptist Convention.

The Yoruba area contains a large Anglican population, while Igboland is predominantly Catholic and the Edo area is predominantly Assemblies of God which was introduced into Nigeria by Gabrial Ojemekele Oyakhilome and his associates at Enugu.

Across Yorubaland in the west many people are adherents to Yorubo/Irunmole spirituality with its philosophy of divine destiny that all can become Orisha (*ori*, spiritual head; *sha*, is chosen: to be one with Olodumare (*oni odu*, the God source of all energy; *ma re*, enlighthens/triumphs). Across Nigeria, there is a growing tendency to abandon Christianity and Islam in favour of traditional religions.

Other minority religious and spiritual groups in Nigeria include Hinduism, Judaism, The Bahai Faith, and Chrislam (a syncretic faith melding elements of Christianity and Islam). Further, Nigeria has become an African hub for the Grail Movement and the Hare Krishnas.

Cuisine

Nigerian cuisine, like West African cuisine in general, is known for its richness and variety. Many different spices, herbs and flavourings are used in conjunction with palm oil or groundnut oil to create deeply flavoured sauces and soups often made very hot with chili peppers. Nigerian feasts are colourful and lavish, while aromatic market and roadside snacks cooked on barbecues or fried in oil are plentiful and varied.

Sport

Soccer is Nigeria's national sport and the country has its own Premier League of football. Nigeria's national football team, known as the Super Eagles, has made the World Cup on four occasions 1994, 1998, 2002, and most recently in 2010. They won the African Cup of Nations in 1980 and 1994, and also hosted the Junior World Cup. They won the gold medal for football in the 1996 Summer Olympics (in which they beat Argentina).

The nation's cadet team to Japan '93 produced some international players notably Nwankwo Kanu, a two-time African Footballer of the year who won the European Champions League with Ajax Amsterdam and later played with Inter Milan (Italy), Arsenal FC (London, UK), West Bromwich Albion (UK) and Portsmouth F.C. (UK). Other players that graduated from the Junior teams are Celestine Babayaro (of Newcastle United, UK), Wilson Oruma and Taye Taiwo (of Marseille, France).

According to the official May 2010 FIFA World Rankings, Nigeria was the second top-ranked football nation in Africa and the 21st highest in the world. Nigeria is also involved in other sports such as basketball, cricket and track and field. Boxing is also an important sport in Nigeria; Dick Tiger and Samuel Peter are both former World Champions.

Societal Issues

Despite its vast government revenue from the mining of petroleum, Nigeria is faced by a number of societal issues due primarily to a history of inefficiency in its governance.

Human Rights

Nigeria's human rights record remains poor and government officials at all levels continue to commit serious abuses.

According to the U.S. Department of State, the most significant human rights problems are: extrajudicial killings and use of excessive force by security forces; impunity for abuses by security forces; arbitrary arrests; prolonged pretrial detention; judicial corruption and executive influence on the judiciary; rape, torture and other cruel, inhuman or degrading treatment of prisoners, detainees and suspects; harsh and life threatening prison and detention centre conditions; human trafficking for the purpose of prostitution and forced labour; societal violence and vigilante killings; child labour, child abuse and child sexual exploitation; female genital mutilation (FGM); domestic violence; discrimination based on sex, ethnicity, region and religion; restrictions on freedom of assembly, movement, press, speech and religion; infringement of privacy rights; and the abridgement of the right of citizens to change the government.

Under the Sharia penal code that applies to Muslims in twelve northern states, offences such as alcohol consumption, homosexuality, infidelity and theft carry harsh sentences, including amputation, lashing, stoning and long prison terms.

Strife and Sectarian Violence

Because of its multitude of diverse, sometimes competing ethno-linguistic groups, Nigeria prior to independence has been faced with sectarian tensions and violence. This is particularly a major issue in the oil-producing Niger Delta region, where both state and civilian forces employ varying methods of coercion in attempts gain control over regional petroleum resources. Some of the ethnic groups like the Ogoni, have experienced severe environmental degradation due to petroleum extraction.

Since the end of the civil war in 1970, some ethnic violence has persisted. There has subsequently been a period of relative harmony since the Federal Government introduced tough new measures against religious violence in all affected parts of the country. In 2002, organizers of the Miss World Pageant were forced to move the pageant from the Nigerian capital, Abuja, to London in the wake of violent protests in the Northern part of the country that left more than 100 people dead and over 500 injured. The rioting erupted after Muslims in the country reacted in anger to comments made by a newspaper reporter. Rioters in Kaduna killed an estimated 105 men, women, and children with a further 521 injured taken to hospital.

Health Issues

Nigeria has been reorganizing its health system since the Bamako Initiative of 1987 formally promoted community-based methods of increasing accessibility of drugs and health care services to the population, in part by implementing user fees. The new strategy dramatically increased accessibility through community-based health-care reform, resulting in more efficient and equitable provision of services. A comprehensive approach strategy was extended to all areas of health care, with subsequent improvement in the health care indicators and improvement in health care efficiency and cost.

The Nigerian health care system is continuously faced with a shortage of doctors known as 'brain drain' due to the fact that many highly skilled Nigerian doctors emigrate to North America and Europe. In 1995, it was estimated that 21,000 Nigerian doctors were practicing in the United States alone, which about the same as the number of doctors working in the Nigerian public service. Retaining these expensively trained professionals has been identified as one of the goals of the government.

Education

Nigeria provides free, government-supported education, but attendance is not compulsory at any level, and certain groups, such as nomads and the handicapped, are under-served. The education system consists of six years of primary school, three years of junior secondary school, three years of senior secondary school, and four years of university education leading to a bachelor's degree. The rate of secondary school attendance is 32 percent for males and 27 percent for females. In 2004 the Nigerian National Planning Commission described the country's education system as "dysfunctional." Reasons for this characterization included decaying institutions and ill-prepared graduates.

Crime

Nigeria is home to a substantial network of organized crime, active especially in drug trafficking. Nigerian criminal groups are heavily involved in drug trafficking, shipping heroin from Asian countries to Europe and America; and cocaine from South America to Europe and South Africa.. The various Nigerian Confraternities or "campus cults" are active in both organized crime and in political violence as well as providing a network of corruption within Nigeria. As confraternities have extensive connections with political and military figures, they offer excellent alumni networking opportunities. The Supreme Vikings Confraternity, for example, boasts that twelve members of the Rivers State House of Assembly are cult members. On lower levels of society, there are the "area boys", organized gangs mostly active in Lagos who specialize in mugging and small-scale drug dealing. According to official statistics, gang violence in Lagos resulted in 273 civilians and 84 policemen killed in the period of August 2000 to May 2001.

Internationally, Nigeria is infamous for a crime dubbed *419*, a type of advance fee fraud (named after Section 419 of the Nigerian Penal Code) along with the "Nigerian scam", a form of confidence trick practiced by individuals and criminal syndicates. In 2003, the Nigerian Economic and Financial Crimes Commission (or EFCC) was created to combat this and other forms of organized financial crime.

There is also some Piracy in Nigeria, with attacks mainly directed at smaller ships shuttling employees and materials belonging to the oil companies with any involvement in oil exploration in the Niger Delta. From January 1, 2007 to October 15, 2007, twenty-six pirate attacks were recorded.

8

Senegal

Senegal, officially the Republic of Senegal, is a country south of the Senegal River in western Africa. It owes its name to the river that borders it to the East and North and that originates from the Fouta Djallon in Guinea. Senegal is externally bounded by the Atlantic Ocean to the west, Mauritania to the north, Mali to the east, and Guinea and Guinea-Bissau to the south; internally it almost completely surrounds The Gambia, namely on the north, east and south, exempting Gambia's short Atlantic Ocean coastline. Senegal covers a land area of almost 197,000 km², and has an estimated population of about 14 million. The climate is tropical with two seasons: the dry season and the rainy season.

Dakar, the capital city of Senegal, is located at the western most tip of the country on the Cap-Vert peninsula. About 300 miles off the coast, in the Atlantic Ocean, lie the Cape Verde Islands. During colonial times, numerous trading Counters, belonging to various colonial empires were established along the coast. The town of St Louis became the capital of French Western Africa (Afrique Occidentale Francaise, or AOF) before it was moved to Dakar in 1902. Dakar later became its capital in 1960 at the time of independence from France.

Etymology

The origin of the name Senegal is controversial. One possible source is in 1850 in the abbot David Boilat's *Esquisses senegalaises* ("Senegalese Sketches"), one sees a deformation of the wolof phrase *sunu gaal*, which means "our canoe". This is the version most often relayed in the media. This theory has been contested since the 1960s, and other etymologies were advanced, for example that it is connected to the name of a Berber tribe from the Sahara, the Zenaga. The scientific debate tends to favour the Zenaga theory today.

History

Archaeological findings throughout the area indicate that Senegal was inhabited in prehistoric times.

Eastern Senegal was once part of the Empire of Ghana. It was founded by the Tukulor in the middle valley of the Senegal River. Islam, the dominant religion in Senegal, first came to the region in the 11th century. In the 13th and 14th centuries, the area came under the influence of the empires to the east; the Jolof Empire of Senegal also was founded during this time. In the Senegambia region, between 1300 and 1900, close to one-third of the population was enslaved. Various European powers—Portugal, the Netherlands, and Great Britain—competed for trade in the area from the 15th century onward, until in 1677, France ended up in possession of what had become a minor slave trade departure point—the island of Goree next to modern Dakar, used as a base to purchase slaves from the warring chiefdoms on the mainland.

The first kingdoms were created around the 7th century, the Tekrour, the Namandirou kingdom and then the Djolof with distant ties to the Ghana empire. In the 14th century the Djolof kingdom became a powerful empire having regrouped the Cayor, the Baol, the Sine and Saloum, the Waalo, the Fouta-Toro and the Bambouk kingdoms. The empire was founded by Ndiadiane N'diaye who was able to form a coalition with many ethnicities but collapsed around 1549 with the defeat and killing of Lele Fouli Fak by Amari Ngone Sobel Fall. French colonialists progressively invaded and took over all kingdoms under their governor Louis Faidherbe. Islam was introduced in Senegal between the 8th and 9th century by Arab merchants. They peacefully converted the Toucouleurs and Sarakholles who in turn propagated it. Later on, in the 11th century, the Almoravids, with the help of the Toucouleurs used Jihad as a mean of conversion. This movement faced resistance from ethnicities of traditional religion and caused them to moved away further in the country (Sineisties) and to the South(Casamance). Eventually, Arabs won a peaceful conversion thanks to the intervention of leaders like Cheikh Ahmadou Bamba,El Hadj Malick Sy, and Seydina Limamou Laye who were able to convince their followers. They saw Islam as a way to unite and fight against colonial power. The populations were getting weary of repeated jihads and forced colonization. Europeans missionaries introduced Christianity to the Sine and Casamance in the 19th century. An emblematic figure of Casamance is Aline Sitoe Diatta, a woman who led the resistance movement against European colonialists.

It was only in the 1850s that the French began to expand onto the Senegalese mainland (by now rid of slavery and promoting abolitionist doctrine), adding native chiefdoms such as Waalo, Cayor, Baol, and Jolof. Senegalese chiefs' resistance to the French expansion and curtailing of their lucrative slave trade was led in part by Lat-Dior, Damel (great chief) of Cayor.

In January 1959 Senegal and the French Sudan merged to form the Mali Federation, which became fully independent on 20 June 1960, as a result of the independence and the transfer of power agreement signed with France on 4 April 1960. Due to internal political difficulties, the Federation broke up on August 20. Senegal and French Sudan (renamed the Republic of Mali) proclaimed independence. Leopold Senghor was proclaimed Senegal's first president in September 1960. Senghor was a very well read man, educated in France. He was a poet, a philosopher and personally drafted the Senegalese national anthem, "Pincez tous vos koras frappez les balafons". As such he was not really a politician but was handed the presidency by the French authorities who saw in him a brilliant and peaceful man and not a revolutionary like Ahmed Sekou Toure of the neighbouring Guinea

Later after the breakup of the Mali Federation, President Senghor and Prime Minister Mamadou Dia governed together under a parliamentary system. Senghor always feared his Prime Minister who was a very charismatic figure and a hard liner. In December 1962 he accused him of an attempted coup and Dia was wrongfully convicted of treason and briefly jailed.

Senegal adopted a new constitution that consolidated the president's power. In 2006, the current president Abdoulaye Wade vacated the conviction and bestowed upon him a Medal of Honour. In 1980 President Senghor decided to retire from politics, and he handed power over in 1981 to his handpicked successor, Abdou Diouf. Mamadou Dia ran for re-election in 1983 against Abdou Diouf but lost. Senghor moved to France where he later died at the age of 96 having been married to a French woman.

Senegal joined with The Gambia to form the nominal confederation of Senegambia on 1 February 1982. However, the union was dissolved in 1989. Despite peace talks, a southern separatist group in the Casamance region had clashed sporadically with government forces since 1982. Senegal has had a long history of participating in international peacekeeping.

Abdou Diouf was president between 1981 and 2000. He encouraged broader political participation, reduced government involvement in the economy, and widened Senegal's diplomatic engagements, particularly with other developing nations. Domestic politics on occasion spilled over into street violence, border tensions, and a violent separatist movement in the southern region of the Casamance. Nevertheless, Senegal's commitment to democracy and human rights strengthened. Diouf served four terms as president.

In the presidential election of 1999, opposition leader Abdoulaye Wade defeated Diouf in an election deemed free and fair by international observers. Senegal experienced its second peaceful transition of power, and its first from one political party to another. On 30 December 2004 President Abdoulaye Wade announced that he would sign a peace treaty with the separatist group in the Casamance region. This, however, has yet to be implemented. There was a round of talks in 2005, but the results did not yet yield a resolution.

Politics

Senegal is a republic with a presidency; the president is elected every five years as of 2001, previously being seven years, by adult votes. The current president is Abdoulaye Wade, re-elected in March 2007.

Senegal has more than 80 political parties. The bicameral parliament consists of the National Assembly, which has 120 seats, and the Senate, which has 100 seats and was reinstituted in 2007. An independent judiciary also exists in Senegal. The nation's highest courts that deal with business issues are the constitutional council and the court of justice, members of which are named by the president.

Currently Senegal has a democratic political culture, being one of the more successful post-colonial democratic transitions in Africa. Local administrators are appointed by, and responsible to, the president. The marabouts, religious leaders of the various Senegalese Muslim brotherhoods, also exercise a strong political influence in the country. In 2009, however, Freedom House downgraded Senegal's status from 'Free' to 'Partially Free', based on increased centralisation of power in the executive.

In 2008, Senegal finished in 10th position on the Ibrahim Index of African Governance. The Ibrahim Index is a comprehensive measure of sub-Saharan African governance, based on a number of different variables which reflect the success with which governments deliver essential political goods to its citizens. In 2009, Senegal's ranking slipped substantially, to 17th but rose to 3rd out of the original 53 countries in 2008 place; however, this is partially accounted for by the addition of Northern African nations to the rankings.

Geography

Senegal is located on the west of the African continent. The Senegalese landscape consists mainly of the rolling sandy plains of the western Sahel which rise to foothills in the southeast. Here is also found Senegal's highest point, an otherwise unnamed feature near Nepen Diakha at 584 m (1,916 ft). The northern border is formed by the Senegal River, other rivers include the Gambia and Casamance Rivers. The capital Dakar lies on the Cap-Vert peninsula, the western most point of continental Africa.

The Cape Verde islands lie some 560 kilometres (348 mi) off the Senegalese coast, but Cap Vert ("Cape Green") is a maritime placemark, set at the foot of "Les Mammelles", a 105-metre (344 ft) cliff resting at one end of the Cap Vert peninsula onto which is settled Senegal's capital Dakar, and 1 kilometre (0.62 mi) south of the "Pointe des Almadies", the western-most point in Africa.

Climate

The local climate is tropical with well-defined dry and humid seasons that result from northeast winter winds and southwest summer winds. Dakar's annual rainfall of about 600 mm (23.6 in) occurs between June and October when maximum temperatures average 27 °C (80.6 °F); December to February minimum temperatures are about 17 °C (62.6 °F). Interior temperatures can be substantially higher than along the coast, and rainfall increases substantially farther south, exceeding 1,500 mm (59.1 in) annually in some areas. The far interior of the country, in the region of Tambacounda, particularly on the border of Mali, temperatures can reach as high as 54 °C (129.2 °F).

Administrative Divisions

Senegal is subdivided into 14 regions, each administered by a *Conseil Regional* (Regional Council) elected by population weight at the *Arrondissement* level. The country is further subdivided by 45 *Departements*, 103 *Arrondissements* (neither of which have administrative function) and by *Collectivites Locales*, which elect administrative officers.

Regional capitals have the same name as their respective regions:

- Dakar
- Diourbel
- Fatick
- Kaffrine
- Kaolack
- Kedougou
- Kolda
- Louga
- Matam
- Saint-Louis
- Sedhiou

- Tambacounda
- Thies
- Ziguinchor.

Major Cities

Senegal's capital of Dakar is by far the largest city in Senegal, with over two million residents. The second most populous city is Touba, a *de jure Communaute rurale* (rural community), with half a million.

City	*Population (2005)*
Dakar (Dakar proper, Guediawaye, and Pikine)	2,145,193
Touba (Touba Mosquee)	475,755
Thies	240,152
Kaolack	181,035
M'Bour	170,875
Saint-Louis	165,038
Rufisque	154,975
Ziguinchor	153,456

Economy

Predominantly rural and with limited natural resources, the Economy of Senegal gains most of its foreign exchange from fish, phosphates, groundnuts, tourism, and services. Its agricultural sector is highly vulnerable to variations in rainfall and changes in world commodity prices. Dakar, as the former capital of French West Africa, is also home to banks and other institutions which serve all of Francophone West Africa, and is a hub for shipping and transport in the region. It also boasts one of the best developed tourist industries in Africa. Senegal still depends heavily on foreign assistance, which in 2000 represented about 32% of overall government spending—including both current expenditures and capital investments—or CFA 270.8 billion (U.S.$ 361.0 million). Senegal is a member of the WTO.

IMF and 1990s Economic Reforms

Since the January 1994 CFA franc devaluation, the International Monetary Fund (IMF), the World Bank, and other multilateral and creditors have been supporting the

Government of Senegal's structural and sectoral adjustment programs. The broad objectives of the program have been to facilitate growth and development by reducing the role of government in the economy, improving public sector management, enhancing incentives for the private sector, and reducing poverty.

In January 1994, Senegal undertook a bold and ambitious economic reform program with the support of the international donor community. This reform began with a 50% devaluation of Senegal's currency, the CFA franc, which was linked at a fixed rate to the French franc. Government price controls and subsidies have been steadily dismantled as another economic reform. However, this devaluation had severe social consequences, because most of the essentials goods were imported. Overnight, the price of goods such as milk, rice, fertilizer and machinery doubled. As a result, the country suffered a large exodus, with many of the most educated people and those who could afford it choosing to leave the country. After an economic contraction of 2.1% in 1993, Senegal made an important turnaround, thanks to the reform program, with a growth in GDP averaging over 5% annually during 1995-2004. Annual inflation had been pushed down to the low single digits. As a member of the West African Economic and Monetary Union (WAEMU), Senegal is working toward greater regional integration with a unified external tariff and a more stable monetary policy. Senegal still relies heavily upon outside donor assistance, however. Under the IMF's Highly Indebted Poor Countries debt relief program, Senegal will benefit from eradication of two-thirds of its bilateral, multilateral, and private sector debt, contingent on the completion of privatization program proposed by the government and approved by the IMF

Current State of Economy

External Trade and Investment

The fishing sector has replaced the groundnut sector as Senegal's export leader. Its export earnings reached U.S.$239 million in 2000. The industrial fishing operations struggle with high costs, and Senegalese tuna is rapidly losing the French market to more efficient Asian competitors.

Phosphate production, the second major foreign exchange earner, has been steady at about US $95 million. Exports of peanut products reached US $79 million in 2000 and represented 11% of total export earnings. Receipts from tourism, the fourth major foreign exchange earner, have picked up since the January 1994 devaluation. In 2000, some 500,000 tourists visited Senegal, earning the country $120 million.

Senegal's new *Agency for the Promotion of Investment* (APIX) plays a pivotal role in the government's foreign investment program. Its objective is to increase the investment rate from its current level of 20.6% to 30%. Currently, there are no restrictions on the transfer or repatriation of capital and income earned, or investment financed with convertible foreign exchange. Direct US investment in Senegal remains about US $38 million, mainly in petroleum marketing, pharmaceuticals manufacturing, chemicals, and banking. Economic assistance, about US $350 million a year, comes largely from France, the IMF, the World Bank, and the United States. Canada, Italy, Japan, and Germany also provide assistance.

Senegal has well-developed though costly port facilities, a major international airport serving 23 international airlines, and direct and expanding telecommunications links with major world centres.

Indebtedness

With an external debt of US $2,495 million, and with its economic reform program on track, Senegal qualified for the multilateral debt relief initiative for Heavily Indebted Poor Countries (HIPC). Progress on structural reforms is on track, but the pace of reforms remains slow, as delays occur in implementing a number of measures on the privatization program, good governance issues, and the promotion of private sector activity. However, macroeconomic indicators show that Senegal turned in a respectable performance in meeting IMF targets in 2000: annual GDP growth increased to 5.7%, compared to 5.1% in 1999. Inflation was reported to be 0.7% compared to 0.8% in 1999, and the current account deficit (excluding transfers) was held at less than 6% of GDP.

Notable Senegalese Companies

Large Senegalese companies include Senelec, the national power company; Air Senegal International airlines; Sonatel, the principal telecommunications provider in Senegal; and Industrie chimique du Senegal, a chemical processing company.

Trade Unions

Senegalese trade unions include The National Confederation of Senegalese Workers (CNTS) and it's affiliate the Dakar Dem Dikk Workers Democratic Union (Dakar Public Transport workers), The Democratic Union of Senegalese Workers (UTDS), The General Confederation Of Democratic Workers Of Senegal (CGTDS) and the National Union of Autonomous Trade Unions of Senegal (UNSAS).

Stock Exchange

Senegal's corporations are included in the Bourse Regionale des Valeurs Mobilieres SA (BRVM), a regional stock exchange serving the following eight West African countries, and located in Abidjan, Cote d'Ivoire.

Demographics

Senegal has a population of over 12.5 million, about 42 percent of whom live in rural areas. Density in these areas varies from about 77 inhabitants per square kilometre (199/sq mi) in the west-central region to 2 inhabitants per square kilometre (5/sq mi) in the arid eastern section.

According to the *World Refugee Survey 2008*, published by the U.S. Committee for Refugees and Immigrants, Senegal has a population of refugees and asylum seekers numbering approximately 23,800 in 2007. The majority of this population (20,200) is from Mauritania. Refugees live in N'dioum, Dodel, and small settlements along the Senegal River valley.

Ethnicity

Senegal has a wide variety of ethnic groups and, as in most West African countries, several languages are widely spoken. The Wolof are the largest single ethnic group in Senegal at 43 percent; the Peul and Toucouleur (also known as Halpulaar, Fulbe or Fula) (24 percent) are the second biggest group, followed by others that include the Serer (15 percent), Lebou (10 percent), Jola (4 percent), Mandinka (3 percent), Maures or Naarkajors, Soninke, Bassari and many smaller communities (9 percent).

About 50,000 Europeans (mostly French) and Lebanese as well as smaller numbers of Mauritanians and Moroccans reside in Senegal, mainly in the cities. The majority of Lebanese work in commerce. Also located primarily in urban settings are small Vietnamese communities as well as a growing number of Chinese immigrant traders, each numbering perhaps a few hundred people. There are also tens of thousands of Mauritanian refugees in Senegal, primarily in the country's north.

From the time of earliest contact between Europeans and Africans along the coast of Senegal, particularly after the establishment of coastal trading posts during the fifteenth century, communities of mixed African and European (mostly French and Portuguese) origin have thrived. Cape Verdean migrants and their descendants living in urban areas and in the Casamance region represent another recognized community of mixed African and European background.

French is the official language, used regularly by a minority of Senegalese educated in a system styled upon the colonial-era schools of French origin (Koranic schools are even more popular, but Arabic is not widely spoken outside of this context of recitation). Most people also speak their own ethnic language while, especially in Dakar, Wolof is the lingua franca. Pulaar is spoken by the Peuls and Toucouleur.

Portuguese Creole is a prominent minority language in Ziguinchor, regional capital of the Casamance, where some residents speak Kriol, primarily spoken in Guinea-Bissau. Cape Verdeans speak their native creole, Cape Verdean Creole, and standard Portuguese.

Health

Public expenditure on health was at 2.4 % of the GDP in 2004, whereas private expenditure was at 3.5 %. Health expenditure was at US$ 72 (PPP) per capita in 2004. The fertility rate was at about 5.2 in the early 2000s. There were 6 physicians per 100,000 persons in the early 2000s. Infant mortality was at 77 per 1,000 live births in 2005.

Religion

Islam is the predominant religion, practiced by approximately 90 percent of the country's population; the Christian community, at 10 percent of the population, includes Roman Catholics and diverse Protestant denominations. There is also a 1 percent population who maintain animism in their beliefs, particularly in the southeastern region of the country.

Islam

Islamic communities are generally organized around one of several Islamic Sufi orders or brotherhoods, headed by a *khalif* (*xaliifa* in Wolof, from Arabic *khalîfa*), who is usually a direct descendant of the group's founder. The two largest and most prominent Sufi orders in Senegal are the Tijaniyya, whose largest sub-groups are based in the cities of Tivaouane and Kaolack, and the Murîdiyya (Murid), based in the city of Touba.

The Halpulaar, a widespread ethnic group found along the Sahel from Chad to Senegal, representing 20 percent of the Senegalese population, were the first to convert to Islam. The Halpulaar, composed of various Fula people groups, named *Peuls* and *Toucouleurs* in Senegal.

Many of the *Toucouleurs*, or sedentary Halpulaar of the Senegal River Valley in the north, converted to Islam around a millennium ago and later contributed to Islam's propagation throughout Senegal. Most communities south of the Senegal River Valley, however, were not thoroughly Islamized until the nineteenth and early twentieth centuries. During the mid-19th century, Islam became a banner of resistance against the traditional aristocracies and French colonialism, and Tijani leaders Al-Hajj Umar Tall and Mabba Jaxu Ba established short-lived but influential Islamic states but were both killed in battle and their territories then annexed by the French.

The spread of formal Quranic school (called *daara* in Wolof) during the colonial period increased largely through the effort of the Tijaniyya. In Murid communities, which place more emphasis on the work ethic than on literary Quranic studies, the term *daara* often applies to work groups devoted to working for a religious leader. Other Islamic groups include the much older Qadiriyya order and the Senegalese Laayeen order, which is prominent among the coastal Lebu. Today, most Senegalese children study at *daaras* for several years, memorizing as much of the Qur'an as they can. Some of them continue their religious studies at informal Arabic schools (*majlis*) or at the growing number of private Arabic schools and publicly funded Franco-Arabic schools. A modern messianic sect in Islam, the Ahmadiyya Muslim Community is also present in the country.

Christianity

About 7% of the population of Senegal adheres to Christianity. Small Roman Catholic communities are mainly found in coastal Serer, Jola, Mankanya and Balant populations, and in eastern Senegal among the Bassari and Coniagui.

The Protestant churches are mainly attended by immigrants but during the second half of the twentieth century Protestant churches led by Senegales leaders from different ethnic groups have evolved. In Dakar Catholic and Protestant rites are practiced by the Lebanese, Capeverdian, European, and American immigrant populations, and among certain Africans of other countries as well as by the Senegalese themselves. Although Islam is Senegal's majority religion, Senegal's first president, Leopold Sedar Senghor, was a Catholic Serer.

Other Religions

Animism, once widely practiced, has declined in Senegal in recent decades, though some Muslims and Christians incorporate elements of animism in their worship. There are small numbers of adherents of Judaism and Buddhism. Judaism is followed by members of several ethnic groups, while Buddhism is followed by a number of Vietnamese.

Bahai Faith

The Bahai Faith in Senegal was established after 'Abdul-Baha, the son of the founder of the religion, mentioned Africa as a place the religion should be more broadly visited by Bahais. The first Bahais to set foot in the territory of French West Africa that would become Senegal arrived in 1953. The first Bahai Local Spiritual Assembly of Senegal was elected in 1966 in Dakar. In 1975 the Bahai community elected the first National Spiritual Assembly of Senegal. The most recent estimate, by the Association of Religion Data Archives in a 2005 report details the population of Senegalese Bahais at 22,000.

Culture

Senegal's musical heritage is better known than that of most African countries, due to the popularity of mbalax, which is a form of Wolof percussive; it has been popularized by Youssou N'Dour. Sabar drumming is especially popular. The sabar is mostly used in special celebrations like weddings. Another instrument, the tama, is used in more ethnic groups. Other popular Senegalese musicians are Ismael Lo, Orchestra Baobab, Baba Maal, Thione Seck, Akon, Viviane, Titi, and Pape Diouf.

Education

Articles 21 and 22 of the Constitution adopted in January 2001 guarantee access to education for all children. Education is compulsory and free up to the age of 16. The Ministry of Labour has indicated that the public school system is unable to cope with the number of children that must enroll each year. Illiteracy is high, particularly among women. The net primary enrollment rate was 69 % in 2005. Public expenditure on education was 5.4 % of the 2002–2005 GDP.

9

Liberia

Liberia, is a country on the west coast of Africa, bordered by Sierra Leone, Guinea, Cote d'Ivoire, and the Atlantic Ocean. As of the 2008 Census, the nation is home to 3,476,608 people and covers 111,369 square kilometres (43,000 sq mi).

Liberia's capital is Monrovia. Liberia has a hot equatorial climate with most rainfall arriving in summer with harsh harmattan winds in the dry season. Liberia's populated Pepper Coast is composed of mostly mangrove forests while the sparsely populated inland is forested, later opening to a plateau of drier grasslands.

The history of Liberia is unique among African nations because of its relationship with the United States. It is one of the few countries in Africa, and the only country in West Africa, without roots in the European Scramble for Africa. It was founded and colonized by freed American slaves with the help of a private organization called the American Colonization Society in 1821-1822, on the premise that former American slaves would have greater freedom and equality there.

Slaves freed from slave ships were also sent there instead of being repatriated to their countries of origin. These colonists formed an elite group in Liberian society, and, in 1847, they founded the Republic of Liberia, establishing a government modelled on that of the United States, naming Monrovia, their capital city, after James Monroe, the fifth president of the United States and a prominent supporter of the colonization. A military-led coup in 1980 overthrew then-president William

R. Tolbert, which marked the beginning of a period of instability that eventually led to two civil wars that left hundreds of thousands of people dead and devastated the country's economy. Today, Liberia is recovering from the lingering effects of the civil war and related economic dislocation. Statistics indicate that about 85% of the population live on less than $1.25 a day.

Etymology

The name Liberia denotes "liberty". The newly arrived settlers formed a new ethnic group called the Americo-Liberians. However, this introduction of a new ethnic mix resulted in ethnic tensions with the sixteen other main ethnicities already residing in Liberia. From the 1500s until 1822, European explorers and traders had multiple names for Liberia, varying by language.

During the spice trade, in non-English speaking Europe, Liberia was called the Malaguetta Coast or Pepper Coast in English. It earned its name from the melegueta pepper found in rural Liberia that was dubbed the "Grains of paradise" since it was a rare spice in high demand throughout continental Europe. In late 18th century English explorers referred to the country as the Windward Coast because of notoriously unnavigable, choppy waters off the coast of Cape Palmas at the tip of Southern Liberia that were difficult for European ships to sail through.

History

Indigenous Peoples of West Africa

Anthropological and archeological research shows the region of Liberia was inhabited at least as far back as the 12th century, perhaps earlier. Mende-speaking people expanded westward, forcing many smaller ethnic groups southward towards the Atlantic ocean. The Days, Bassa, Kru, Gola and Kissi were some of the earliest recorded arrivals. This influx was compounded during the ancient decline of the Western Sudanic Mali Empire in 1375 and later in 1591 with the Songhai Empire. Additionally, inland regions underwent desertification, and inhabitants were pressured to move to the wetter Pepper Coast. These new inhabitants brought skills such as cotton spinning, cloth weaving, iron smelting, rice and sorghum cultivation, and social and political institutions from the Mali and Songhai Empires.

Shortly after the Manes conquered the region, there was a migration of the Vai people into the region of Grand Cape Mount. The Vai were part of the Mali Empire who were forced to migrate when the empire collapsed in the 14th century. The Vai chose to migrate to the coastal region. The ethnic Kru opposed the influx of Vai. An alliance of the Manes and Kru was able to stop further influx of Vai, but the Vai remained in the Grand Cape Mount region (where the city of Robertsport is now located).

People of the Littoral coast built canoes and traded with other West Africans from Cap-Vert to the Gold Coast. Later European traders would barter various commodities and goods with local people, sometimes hoisting their canoes aboard. When the Kru began trading with Europeans, they initially traded in commodities, but later they actively participated in the African slave trade.

Kru laborers left their territory to work as paid laborers on plantations and in construction. Some even worked building the Suez and Panama Canals.

Another ethnic group in the area was the Glebo. The Glebo were driven, as a result of the Manes invasion, to migrate to the coast of what later became Liberia.

Between 1461 and late 17th century, Portuguese, Dutch and British traders had contacts and trading posts in what became Liberia. The Portuguese had named the area *Costa da Pimenta* (meaning Pepper Coast), later translated as Grain Coast, because of the abundance of grains of melegueta pepper.

Settlers from the United States

In 1822, the American Colonization Society (A.C.S.), working to "repatriate" black Americans to greater freedom in Africa, established Liberia as a place to send people who were formerly enslaved. This movement of black people by the A.C.S. had broad support nationwide among white people in the United States, including politicians such as Henry Clay and James Monroe. They believed this was preferable to emancipation of slaves in the United States. Clay said, because of "unconquerable prejudice resulting from their colour, they never could amalgamate with the free whites of this country. It was desirable, therefore, as it respected them, and the residue of the population of the country, to drain them off." The institution of slavery in the U.S. had grown, reaching almost four million slaves by the mid 1800s. Some free African Americans chose to emigrate to Liberia. The immigrants became known as Americo-Liberians. Many present-day Liberians trace their ancestry to them. On July 26, 1847, Americo-Liberian settlers declared independence of the Republic of Liberia.

The settlers regarded Africa as a "Promised Land," but they did not integrate into African society. Once in Africa, they referred to themselves as Americans. They were recognized as such by local Africans and by British colonial authorities in neighbouring Sierra Leone. The symbols of their state — its flag, motto, and seal — and the form of government which they chose reflected their American background and diaspora experience. Ashmun Institute, founded in Pennsylvania in 1854 for the education of black Americans, played an important role in supplying Americo-Liberians with leadership for the new nation. The first graduating class of Ashmun Institute (later renamed Lincoln University in honour of the slain President), James R. Amos, his brother Thomas H. Amos, and Armistead Miller, sailed for Liberia on the brig *Mary C. Stevens* in April 1859 after graduation.

The religious practices, social customs and cultural standards of the Americo-Liberians had their roots in the antebellum American South. These ideals strongly influenced the attitudes of the settlers toward the indigenous African people. The new nation, as they perceived it, was coextensive with the settler community and with those Africans who were assimilated into it. Mutual mistrust and hostility between the "Americans" along the coast and the "Natives" of the interior was a recurrent theme in the country's history. The Americo-Liberian minority worked to dominate the native people, whom they considered savage primitives. The immigrants named the land "Liberia," which in the Romance languages, and in Latin in particular, means "Land of the Free," as an homage to freedom from slavery. Historically, Liberia has enjoyed the support and unofficial cooperation of the United States government. Liberia's government, modelled after that of the U.S., was democratic in structure, if not always in substance. In 1877, the True Whig Party monopolized political power in the country. Competition for office was usually contained within the party, whose nomination virtually ensured election. Two problems confronting successive administrations were pressure from neighbouring colonial powers, Britain and France, and the threat of financial insolvency, both of which challenged the country's sovereignty. Liberia retained its independence during the Scramble for Africa, but lost its claim to extensive territories that were annexed by Britain and France. Economic development was hindered by the decline of markets for Liberian goods in the late 19th century and by indebtedness on a series of loans, payments on which drained the economy.

Mid-20th Century

Two events were particularly important in releasing Liberia from its self-imposed isolation. The first was the grant in 1926 of a large concession to the American-owned Firestone Plantation Company; that move became a first step in the (limited) modernization of the Liberian economy. The second occurred during World War II, when the United States began providing technical and economic assistance that enabled Liberia to make economic progress and introduce social change. Both the Freeport of Monrovia and Roberts International Airport were built by U.S. personnel during World War II.

On April 12, 1980, a successful military coup was staged by a group of noncommissioned army officers led by Master Sergeant Samuel Kanyon Doe. The soldiers were a mixture of the various ethnic groups that claimed marginalization at the hands of the minority Americo-Liberian settlers. In a late-night raid, they killed William R. Tolbert, Jr., who had been president for nine years, in his mansion. Constituting themselves as the People's Redemption Council, Doe and his associates seized control of the government and brought an end to Africa's first republic. Significantly, Doe was the first Liberian head of state who was not a member of the Americo-Liberian elite. Doe favoured authoritarian policies, banning newspapers and outlawing various opposition parties. His tactic was to brand popular opposition parties as "socialist", and therefore illegal according to the Liberian constitution, while allowing less popular minor parties to remain as a token opposition. Unfortunately for Doe, popular support would then tend to realign behind one of these smaller parties, causing them in turn to be labelled "socialist."

In October 1985, Liberia held the first post-coup elections, ostensibly to legitimize Doe's regime. Virtually all international observers agreed that the Liberia Action Party (LAP) led by Jackson Doe (no relation) had won the election by a clear margin. After a week of counting the votes, however, Samuel Doe fired the count officials and replaced them with his own Special Election Committee (SECOM), which announced that Samuel Doe's ruling National Democratic Party of Liberia had won with 50.9% of the vote. In response, on November 12 a counter-coup was launched by Thomas Quiwonkpa, whose soldiers briefly occupied the Executive Mansion and the national radio station, with widespread support throughout the country. Three days later, Quiwonkpa's coup was overthrown. Government repression intensified, as Doe's troops killed more than 2,000 civilians and imprisoned more than 100 opposing politicians, including Jackson Doe and BBC journalist Isaac Bantu.

1989 and 1999 Civil Wars

In late 1989, the First Liberian Civil War began. The harsh dictatorial atmosphere that gripped the country was due largely to Samuel Doe's rule. Americo-Liberian Charles Taylor, with the backing of neighbouring countries such as Burkina Faso and Cote d'Ivoire, entered Nimba County with around 100 men. These fighters quickly gained control of much of the country, thanks to strong support from the local population who were disillusioned with their then government. By then, a new player also emerged: Yormie Prince Johnson (former ally of Taylor) had formed his own army and had gained tremendous support from the Gio and Mano ethnic groups.

In August 1990, the Economic Community Monitoring Group under the Economic Community of West African States (ECOWAS) organized its own military task force to intervene in the crisis. The troops were largely from Nigeria, Guinea and Ghana. On his way out after a meeting, Doe, who was travelling only with his personal staff, was ambushed and captured by members of the Gio Tribe who were loyal to Prince Yormie Johnson. The soldiers took him to Johnson's headquarters in neighbouring Caldwell, tortured and killed him.

By then, Taylor was a prominent warlord and leader of the National Patriotic Front of Liberia. After some prompting from Taylor that the anglophone Nigerians and Ghanaians were opposed to him, Senegalese troops were brought in with some financial support from the United States. But their service was short-lived, after a major confrontation with Taylor's forces in Vahun, Lofa County on 28 May 1992, when six were killed when a crowd of NPFL supporters surrounded their vehicle and demanded they surrender the vehicle and weapons.

By September 1990, Doe's forces controlled only a small area just outside the capital, Monrovia. After Doe's death, and as a condition for the end of the conflict, interim president Amos Sawyer resigned in 1994, handing power to the Council of State. Taylor was elected as President in 1997, after leading a bloody insurgency backed by Libyan President Muammar al-Gaddafi. Taylor's brutal regime targeted several leading opposition and political activists. In 1998, the government sought to assassinate child rights activist Kimmie Weeks for a report he had published on its involvement in the training of child soldiers, which forced him into exile. Taylor's autocratic and dysfunctional government led to the Second Liberian Civil War in 1999.

The conflict intensified in mid-2003, and the fighting moved into Monrovia. An elite rapid response unit of the US Marines known as 'FAST' deployed to the US Embassy to ensure the security and interests of the US. The Marines used US Air Force HH-60 Pave Hawk to airlift non-combatants and foreign nationals to Dakar, Senegal. A hastily assembled force of 1,000 Nigerian troops, the ECOWAS Mission In Liberia (ECOMIL), was airlifted into Liberia on August 15, 2003 to prevent the rebels from overrunning the capital city and committing revenge-inspired war crimes. Meanwhile the US Joint Task Force Liberia commanded from USS Iwo Jima (LHD-7) was offshore, though only 100 of the 2,000 US Marines landed to meet with the ECOMIL force.

A peace movement called Women of Liberia Mass Action for Peace was instrumental to the end of hostilities in Monrovia. Organized by social worker Leymah Gbowee, thousands of Christian and Muslim women staged silent protests and forced a meeting with President Charles Taylor and extracted a promise from him to attend peace talks in Ghana. Gbowee then led a delegation of Liberian women to Ghana to continue to apply pressure on the warring factions during the peace process. They staged a sit in outside of the Presidential Palace, blocking all the doors and windows and preventing anyone from leaving the peace talks without a resolution. The women of Liberia became a political force against violence and against their government. Their actions brought about an agreement during the stalled peace talks. As a result, the women were able to achieve peace in Liberia after a 14-year civil war and later helped bring to power the country's first female head of state. The story is told in the 2008 documentary film *Pray the Devil Back to Hell.*

As the power of the government shrank, and with increasing international and US pressure for him to resign, President Taylor accepted an asylum offer from Nigeria, but vowed: "God willing, I will be back." Some of the ECOMIL troops were subsequently withdrawn and at least two battalions incorporated into the 15,000 strong United Nations Mission in Liberia (UNMIL) peacekeeping force. More than 200,000 people are estimated to have been killed in the civil wars.

Post Civil War

After the exile of Taylor, Gyude Bryant was appointed Chairman of the transitional government in late 2003. Because of failures of the Transitional Government in curbing corruption, Liberia signed onto GEMAP, a novel anti-corruption program. The primary task of the transitional government was to prepare for fair and peaceful democratic elections. With UNMIL troops safeguarding the peace, Liberia successfully conducted presidential elections on October 11, 2005. There were 23 candidates; an early favourite was George Weah, internationally famous footballer, UNICEF Goodwill Ambassador and member of the Kru ethnic group expected to dominate the popular vote. No candidate took the required majority, prompting a run-off election between the top two candidates, Weah and Ellen Johnson Sirleaf. The November 8, 2005 presidential runoff election was won decisively by Sirleaf, a Harvard-trained economist. Both the general election and runoff were marked by peace and order, as thousands of Liberians waited in the Liberian heat to cast their ballots.

Prior to her election as president, Sirleaf was jailed twice during the Doe administration before escaping and going into exile. Upon taking office she became the first elected female head of state in Africa. During her administration President Sirleaf established a Truth and Reconciliation Commission to address crimes committed during the later stages of Liberia's long civil war. Elsewhere, the Special Court for Sierra Leone (a war crimes tribunal) charged former President Charles Taylor with crimes against humanity, violations of the Geneva Conventions and "other serious violations of international humanitarian law". The indictment was issued on March 29, 2006, and he was later extradited from Nigeria to Sierra Leone, but the trial by the Special Court is being held in The Hague, for security reasons.

Politics and Government

Liberia has a dual system of statutory law based on Anglo-American common law for the modern sector and customary unwritten law for the native sector for exclusively rural ethnic communities. Liberia's modern sector has three equal branches of government in the constitution, though in practice the executive branch headed by the President of Liberia is the strongest of the three. The other two branches are the legislative and judicial.

Following the dissolution of the Republican Party in 1876, the True Whig Party dominated the Liberian government until the 1980 coup. Currently, no party has majority control of the legislature. The longest serving president in Liberian history was William Tubman, serving from 1944 until his death in 1971. The shortest term was held by James Skivring Smith, who controlled the government for two months. However, the political process from Liberia's founding in 1847, despite widespread corruption, was very stable until the end of the First Republic in 1980.

The Executive branch of the government is headed by the President. Other parts of the branch are the Cabinet and the Vice President. Presidents are elected to six-year terms and can serve up to two terms in office. The President is both the head of state and the head of the government, and resides at the Executive Mansion in Monrovia.

The Legislature of Liberia is a bicameral body with an upper chamber Senate and the lower chamber House of Representatives. Each county sends two senators to the legislature for a total of 30 senators, while the 64 seats in the House are distributed among the 15 counties based on the number of registered voters, with a minimum of at least two from each county. Senators serve nine-year terms (only six-year terms for junior senators elected in 2005) and members of the House six-year terms. Leadership consists of a Speaker in the House and a President Pro Tempore in the Senate. Liberia's Vice President serves as the President of the Senate. The legislature meets in the capital city of Monrovia.

Liberia's highest judicial authority is the Supreme Court, headed by the Chief Justice. The five justice court holds sessions at the Temple of Justice on Capitol Hill in Monrovia. Members are nominated to the court by the President and are confirmed by the Senate and have lifetime tenure. Under the supreme court are 15 circuit courts, one in each county.

Liberia's former minister of justice has been at the centre of a legal scandal for what amounts to de facto withholding the Liberian law, preventing it from being reprinted and circulated.

Human Rights

Amnesty International pawan summarizes in its Annual Report 2006:

> *"Sporadic outbreaks of violence continued to threaten prospects of peace. Former rebel fighters who should have been disarmed and demobilized protested violently when they did not receive benefits. Slow progress in reforming the police, judiciary and the criminal justice*

system resulted in systematic violations of due process and vigilante violence against criminal suspects. Laws establishing an Independent National Commission on Human Rights and a Truth and Reconciliation Commission were adopted. Over 200,000 internally displaced people and refugees returned to their homes, although disputes over land and property appropriated during the war raised ethnic tensions. UN sanctions on the trade in diamonds and timber were renewed. Those responsible for human rights abuses during the armed conflict continued to enjoy impunity. The UN Security Council gave peacekeeping forces in Liberia powers to arrest former President Taylor and transfer him to the Special Court for Sierra Leone if he should return from Nigeria, where he continued to receive asylum. Liberia made a commitment to abolish capital punishment. A new law on rape, which initially proposed imposition of the death penalty for gang rape, was amended to provide a maximum penalty of life imprisonment."

Former 22nd president Charles Taylor was later captured trying to escape across the border of Cameroon and has been sent to the International Criminal Court in The Hague for trial.

In 2009, the country set up a new court to deal with rape crimes in and around Monrovia, called "Court E." The court operates in camera, shielding the victims who testify from exposure to the accused perpetrator and the rest of the courtroom. The prosecutor operates out of the Sexual and Gender Based Violence Crimes office, which offers victims' support services, but faces many financial, logistical and legal hurdles.

Geography

Liberia is situated in West Africa, bordering the North Atlantic Ocean to the country's southwest. The landscape is characterized by mostly flat to rolling coastal plains that contain mangroves and swamps, which rise to a rolling plateau and low mountains in the northeast. Tropical rainforests cover the hills, while elephant grass and semi-deciduous forests make up the dominant vegetation in the northern sections. The equatorial climate is hot year-round with heavy rainfall from May to

October with a short interlude in mid-July to August. During the winter months of November to March, dry dust-laden harmattan winds blow inland, causing many problems for residents.

Liberia's watershed tends to move in a southwestern pattern towards the sea as new rains move down the forested plateau off the inland mountain range of Guinee Forestiere, in Guinea. Cape Mount near the border with Sierra Leone receives the most precipitation in the nation. The country's main northwestern boundary is traversed by the Mano River while its southeast limits are bounded by the Cavalla River. Liberia's three largest rivers are St. Paul exiting near Monrovia, the river St. John at Buchanan and the Cestos River, all of which flow into the Atlantic. The Cavalla is the longest river in the nation at 515 kilometres (320 mi).

The highest point wholly within Liberia is Mount Wuteve at 1,440 metres (4,724 ft) above sea level in the northwestern Liberia range of the West Africa Mountains and the Guinea Highlands. However, Mount Nimba near Yekepa, is higher at 1,752 metres (5,748 ft) above sea level but is not wholly within Liberia as Nimba shares a border with Guinea and Cote d'Ivoire (Ivory Coast) and is their tallest mountain as well.

Counties and Districts

Liberia is divided into 15 counties, which are subdivided into districts, and further subdivided into *clans*. The oldest counties are Grand Bassa and Montserrado, both founded in 1839 prior to Liberian independence. Gbarpolu is the newest county, created in 2001. Nimba is the largest of the counties in size at 11,551 square kilometres (4,460 sq mi), while Montserrado is the smallest at 1,909 square kilometres (737 sq mi). Montserrado is also the most populous county with 1,144,806 residents as of the 2008 census.

Economy

Liberia is considered one of the worlds poorest countries, with 85% unemployment. Historically, the Liberian economy depended heavily on iron ore and rubber exports, foreign direct investment, and exports of other natural resources, such as timber. Agricultural products include livestock (goats, pigs, cattle) and rice, the staple food. Fish are raised on inland farms and caught along the coast. Other foods are imported to support the population. Electricity is provided by dams and oil-fired plants.

Foreign trade was primarily conducted for the benefit of the Americo-Liberian elite. The 1864 Ports of Entry Act severely restricted trade between foreigners and indigenous Liberians throughout most of Liberia's history. Little foreign direct investment benefited the 95% majority population, who were often subjected to forced labour on foreign concessions. Liberian law often did not protect indigenous Liberians from the extraction of rents and arbitrary taxation, and the majority survived on subsistence farming and low wage work on foreign concessions.

While official export figures for commodities declined during the 1990s' civil war as many investors fled, Liberia's wartime economy featured the exploitation of the region's diamond wealth. The country acted as a major trader in Liberian, Sierra Leonian and Angolan blood diamonds, exporting over $300 million in diamonds annually. This led to a United Nations ban on Liberian diamond exports, which was lifted on April 27, 2007.

Other commodity exports continued during the war, in part due to illicit agreements between Liberia's warlords and foreign concessionaires. Looting and war profiteering destroyed nearly the entire infrastructure of the country, such that the Monrovian capital was without running water and electricity (except for fuel-powered generators) by the time the first elected post-war government began to institute development and reforms in 2006.

Once the hostilities ended, some official exporting and legitimate business activity resumed. For instance, Liberia signed a new deal with steel giant Mittal for the export of iron ore in summer 2005. But, as of mid-2006 Liberia was still dependent on foreign aid, and had a debt of $3.5 billion. As of 2003, Liberia had an estimated 85% unemployment rate, the second highest in the world behind only Nauru.

The Liberia dollar currently trades against the US dollar at a ratio of 65:1. Liberia used the US dollar as its currency from 1943 until 1982. Its external debt ($3.5 billion) is huge compared to its GDP ($2.5 billion/year); it imports approximately $4.839 billion in goods per year, while it exports only about $910 million. Inflation is falling, but still significant (15% in 2003, 4.9% in the 3rd quarter of 2005); interest rates are high, with the average lending rate listed by the Central Bank of Liberia at 17.6% for 3rd quarter 2005 (although the average time deposit rate was only 0.4%, and CD rate only 4.4%, barely keeping pace with inflation).

Liberia is trying to revive its economy post civil war. Various sanctions imposed by UN on diamond and timber exports were removed by 2007. The country has second-largest maritime registry in the world, with some 3000 vessels registered under its flag.

It continues to suffer with poor economic performance due to a fragile security situation, the devastation wrought by its long war, its lack of infrastructure, and necessary human capital to help the country recover from the scourges of conflict and corruption. Liberia has one of the world's largest national registries of ships, due to its status as a "flag of convenience".

Weights and Measures

In 1921, one authority stated, "The metric system is seldom found in Liberia," and according to the CIA Handbook, Liberia primarily uses a non-metric system of units. However, in 2008, the African Development Bank stated that Liberia used the metric system. *Cities of the World* states that Monrovia, the capital, used the metric system. and a report from the University of Tennessee stated that the changeover from English to Metric measures was confusing to coffee and cocoa farmers, that the weighing machines were owned by the buyers and there was no system of weights and measures certification.

However, the 2008 Liberian census used square miles to express population densities and the County Development Agendas of that year were inconsistent, some giving measurements in metric units and some giving them in the older units. The 2009 Annual report of the Ministry of Mines, Lands and Energy used kilometres for road distances but acres for land areas. In 2010, Government press releases have used kilometres for road distances though one report gave the length of a bridge in feet while another press release gave two road distances in kilometres and another road distance in miles. In the measurement of land areas, one press release used acres while an announcement from the Ministry of Agriculture used hectares.

Transport

Railways

Total: 490 km (328 km single track)

Note: in 1989, Liberia had three rail systems owned and operated by foreign steel and financial interests in conjunction with the Liberian Government; one of these, the Lamco Railroad, closed in 1989 after iron ore production ceased; the other two were shut down by the civil war; large sections of the rail lines have been dismantled; approximately 60 km (37 mi) of railroad track was exported for scrap (2001)

Standard gauge: 345 km of 1,435 mm (4 ft 8+^{1}D $_{2}$ in) gauge

Narrow gauge: 145 km of 1,067 mm (3 ft 6 in) gauge

There are two separate systems, with two lines from Monrovia and one line from Buchanan.

Railway Links with Adjacent Countries

- Sierra Leone-no-break-of-gauge 1,067 mm (3 ft 6 in)
  - Sierra Leone
- Guinea-no-break-of-gauge-1,435 mm (4 ft 8+^{1}D $_{2}$ in) & 1,067 mm (3 ft 6 in)
  - Guinea
- Cote d'Ivoire-no-break-of-gauge 1,000 mm (3 ft 3+^{3}D $_{8}$ in)
  - Cote d'Ivoire.

2007

- Talks started with the Mittal Steel company to reopen the former Lamco iron mine and railway.

2006

- UPGRADE FOR BONG-15 June 2006-Negotiations for

upgrading the 87 km Bong Mining Railway from Monrovia are at an advanced stage.

- Ircon (the Indian Railway Construction Corporation), the preferred bidder, hopes to sign a contract worth some $US110 million before the end of June, according to managing director to rehabilitate the 250km line from Buchanan.

Highways

Total: 10,600 km (6,586 mi) (there is major deterioration on all highways due to heavy rains and lack of maintenance)

Paved: 657 km (408 mi)

Unpaved: 9,943 km (6,178 mi) (1996 est.)

When construction and reconstruction of roads in Liberia is complete, the Trans-West African Coastal Highway will cross the country, connecting it to Freetown (Sierra Leone), Abidjan (Cote d'Ivoire), and eventually to 11 other nations of ECOWAS.

Ports and Harbors

- Buchanan-railhead of 1435mm (4 ft 8½ in) gauge for civil war affected iron mine at Nimba
- Greenville
- Harper
- Monrovia.

Merchant Marine

Total: 1,513 ships (1,000 GRT or over) totalling 51,912,244 GRT/79,297,046 metric tons deadweight (DWT)

Ships by type: barge carrier 3, bulk 313, cargo 89, chemical tanker 167, combination bulk 16, combination ore/oil 32, container 318, liquefied gas 99, multi-functional large-load carrier 4, passenger 23, petroleum tanker 302, refrigerated cargo 69, roll on/roll off 20, short-sea passenger 3, specialized tanker 13, vehicle carrier 42

Note: includes some foreign-owned ships registered here as a flag of convenience: Argentina 9, Australia 2, Austria 15, Belgium 9, Brazil 5, Canada 4, Cayman Islands 1, Chile 7, China 39, Croatia 11, Denmark 4, Ecuador 1, Estonia 1, Germany 437, Greece 154, Hong Kong 69, India 5, Indonesia

1, Israel 1, Italy 5, Japan 90, Latvia 20, Man, Isle of 5, Monaco 56, Netherlands 12, New Zealand 1, Nigeria 1, Norway 103, Pakistan 1, Portugal 5, Russia 66, Saudi Arabia 21, Singapore 20, Slovenia 1, South Africa 1, South Korea 10, Spain 2, Sweden 9, Switzerland 17, Taiwan 29, Turkey 3, Ukraine 4, United Arab Emirates 12, United Kingdom 39, United States 113, Uruguay 3, Vietnam 1 (2002 est.)

Airports

Airports-With Paved Runways

Total: 2

Over 3,047 m (10,000 ft): 1

1,524 to 2,437 m (5,000 to 8,000 ft): 1 (2002).

Airports-With Unpaved Runways

Total: 45

1,524 to 2,437 m (5,000 to 8,000 ft): 4

914 to 1,523 m (3,000 to 4,999 ft): 7

Under 914 m (3,000 ft): 34 (2002).

Demographics

As of the 2008 national census, Liberia was home to 3,476,608 people. Of those, 1,118,241 lived in Montserrado County, the most populous county in the country and home to the capital of Monrovia, with the Greater Monrovia district home to 970,824 people. Nimba County is the next most populous county with 462,026 residents. Prior to the 2008 census, the last census had been held in 1984, and it listed the population as 2,101,628. The population of Liberia was 1,016,443 in 1962 and increased to 1,503,368 in 1974.

The population of over 3 million comprises 16 indigenous ethnic groups and various foreign minorities. Indigenous peoples comprise about 95% of the population, the largest of which are the Kpelle in central and western Liberia. Americo-Liberians, who are descendants of African-American settlers, make up 2.5%, and Congo people, descendants of repatriated Congo and Afro-Caribbean slaves who arrived in 1825, make up an estimated 2.5%. There also is a sizable number of Lebanese,

Indians, and other West African nationals who make up a significant part of Liberia's business community. A small minority of Liberians of European descent (estimated at 18,000 in 1999; probably fewer now) reside in the country.

As of 2006, Liberia has the highest population growth rate in the world (4.50% per annum). Similar to its neighbours, it has a large youth population, with half of the population under the age of 18.

Of the population, 40% hold indigenous beliefs, 40% are Christians, and 20% are Muslims.

Health

Life expectancy at birth was at 44.7 in 2005. The fertility rate was at 6.8 births per woman in the early 21st century. Expenditure on health was 22 USD (PPP) in 2004. The infant mortality rate was at 15.7% in 2005 (155.8 deaths for every 1000 live births). The HIV/AIDS prevalence was at 1.7 percent of the adult population as of January, 2009.

Culture

Liberia was traditionally noted for its hospitality, academic institutions, cultural skills, and arts/craft works. Liberia has a long, rich history in textile arts and quilting. The free and former US slaves who emigrated to Liberia brought with them their sewing and quilting skills. The census of 1843 indicated a variety of occupations, including hatter, milliner, seamstress and tailor. Liberia hosted National Fairs in 1857 and 1858 in which prizes were awarded for various needle arts. One of the most well-known Liberian quilters was Martha Ann Ricks, who presented a quilt featuring the famed Liberian coffee tree to Queen Victoria in 1892.

In modern times, Liberian presidents would present quilts as official government gifts. The John F. Kennedy Library and Museum collection includes a cotton quilt by Mrs. Jemima Parker which has portraits of both Liberian president William Tubman and JFK. Zariah Wright-Titus founded the Arthington (Liberia) Women's Self-Help Quilting Club (1987). In the early 1990s, Kathleen Bishop documented examples of appliquéd Liberian quilts. When current Liberian President Ellen Johnson

Sirleaf moved into the Executive Mansion, she reportedly had a Liberian-made quilt installed in her presidential office.

The tallest man-made structure of Africa, the mast of former Paynesville Omega transmitter, is situated in Liberia.

Religion

It is estimated that as much as 40 percent of the population of Liberia practices either Christianity or Christianity combined with elements of traditional indigenous religious beliefs. Approximately 40 percent exclusively practices traditional indigenous religious beliefs. An estimated 20 percent of the population practices Islam. A small percentage is Bahai, Hindu, Sikh, Buddhist, or atheist.

Education

The University of Liberia is the country's largest college and is located in Monrovia. Opened in 1862, it is one of Africa's oldest institutes of higher learning organized upon the western model. Civil war severely damaged the university in the 1990s, but the university has begun to rebuild following the restoration of peace. The school includes six colleges, including a medical school and the nation's only law school, Louis Arthur Grimes School of Law.

Cuttington University was established by the Episcopal Church of the USA (ECUSA) in 1889; its campus is currently located in Suakoko, Bong County (120 miles north of Monrovia). The private school, the oldest private college in Liberia, also holds graduate courses in Monrovia.

According to statistics published by UNESCO for 2004 65% of primary-school age and 24% of secondary-school age children were enrolled in school. This is a significant increase on previous years; the statistics also show substantial numbers of older children going back to earlier school years. On average, children attain 10 years of education, 11 for boys and 8 for girls. Children ages five to eleven are required by law to attend school, though enforcement is lax. A 1912 law required children ages 6 to 16 to attend school. The African Methodist Episcopal University is another fast growing university in the capital.

10

The United Kingdom of Libya

Libya s a country located in North Africa. Bordering the Mediterranean Sea to the north, Libya lies between Egypt to the east, Sudan to the southeast, Chad and Niger to the south, and Algeria and Tunisia to the west.

With an area of almost 1,800,000 square kilometres (694,984 sq mi), Libya is the fourth largest country in Africa by area, and the 17th largest in the world. The capital, Tripoli, is home to 1.7 million of Libya's 5.7 million people. The three traditional parts of the country are Tripolitania, Fezzan, and Cyrenaica. Libya has the highest HDI in Africa and the fourth highest GDP (PPP) per capita in Africa as of 2009, behind Seychelles, Equatorial Guinea and Gabon. These are largely due to its large petroleum reserves and low population.

The flag of Libya consists of a green field with no other characteristics. It is the only national flag in the world with just one colour and no design, insignia, or other details.

Name

The name of the *Libu* Berber tribesmen in Greek was adopted as *Libucs*, Latinised to *Libyes* (with Greek ϖ transcribed as *y* in Latin). The land of the Libu in Doric was *Libua* (or in Attic *Libuc*), Latinised *Libya*. In Classical Greece the term had a broader meaning, encompassing all the continent that later (2nd century BC) came to be known as *Africa*, in antiquity assumed to make up one third of the world's landmass, besides

Europe and Asia. During the Islamic Middle Ages (Ibn Khaldun), the Libu were known as the *Lawata*.

The word *jamahiriyya*, which appears in the full title of the country, is the Arabic term for "republic".

History

Ancient Libya

Archaeological evidence indicates that from as early as 8000 BC, the coastal plain of Ancient Libya was inhabited by a Neolithic people, the Berbers, who were skilled in the domestication of cattle and the cultivation of crops.

Later, the area known in modern times as Libya also was occupied by a series of other peoples, with the Phoenicians, Carthaginians, Greeks, Persian Empire, Romans, Vandals, Arabs, Turks and Byzantines ruling all or part of the area.

Although the Greeks and Romans left ruins at Cyrene, Leptis Magna, and Sabratha, little other evidence remains of these ancient cultures. Some cultural and religious exchanges occurred with the Ancient Egyptians, especially in the northern portion containing the delta of the Nile, that is called Lower Egypt. The prehistoric evidence is fragmentary, but historical records later document continued influences.

Pockets of Berber population remain in modern Libya, but dispersal of Berbers north as far as Ireland and Scandinavia is documented in genetic markers studied by physical anthropologists and dispersal in Africa from the Atlantic coast to the Siwa oasis in Egypt, seems to have followed climatic changes causing increasing desertification. Now the greatest number of Berbers in Africa is in Morocco (about 42% of the population) and in Algeria (about 27% of the population), as well as Tunisia and Libya, but exact statistics are not available;

Phoenicians

The Phoenicians were the first to establish trading posts in Libya, when the merchants of Tyre (in present-day Lebanon) develcped commercial relations with the Berber tribes and made treaties with them to ensure their cooperation in the exploitation of raw materials. By the fifth century BC the greatest of the Phoenician colonies, Carthage, had extended its hegemony across much of North Africa, where a distinctive civilization, known as Punic, came into being. Punic settlements on the Libyan coast included Oea (Tripoli), Libdah (Leptis Magna) and Sabratha. All these were in an area that later was called Tripolis, or "Three Cities". Libya's current-day capital Tripoli takes its name from this. Folklore also talks of the legendary leader Ossama Al-Fitory as a champion of social justice and an astutute military tactician.

Greeks

The Greeks conquered Eastern Libya when, according to tradition, emigrants from the crowded island of Thera were commanded by the oracle at Delphi to seek a new home in North Africa. In 630 BC, they founded the city of Cyrene. Within 200 years, four more important Greek cities were established in the area: Barce (Al Marj); Euhesperides (later Berenice, present-day Benghazi); Teuchira (later Arsinoe, present-day Tukrah); and Apollonia (Susah), the port of Cyrene. Together with Cyrene, they were known as the Pentapolis (Five Cities).

Romans

The Romans unified all three regions of Libya. Tripolitania and Cyrenaica became prosperous Roman provinces and remained so for more than six hundred years. Roman ruins, such as those of Leptis Magna, attest to the vitality of the region during the Roman occupation.

At the time, populous cities and even small towns enjoyed the amenities of urban life consistent with those in Rome. Merchants and artisans from many parts of the Roman world established themselves in North Africa, but the character of the cities of Tripolitania remained decidedly Punic and, in Cyrenaica, Greek.

Under Islam

Libya was conquered by Uqba ibn Nafi in 644 and fully conquered in 655, forming part of the Ummayad Caliphate. This was superseded by the Abbasids in 750, but in practice Libya enjoyed considerable local autonomy under the Aghlabid dynasty. Arab soldiers, spreading their new religion of Islam, entered Cyrenaica in 642 and occupied Tripoli in 643. A succession of Arab and Berber dynasties then controlled what is now Libya. The culture of northwestern Libya developed along with the political units just west of it, while development in the east was strongly influenced by neighbouring Egypt.

Ottoman Turks

The Ottoman Turks conquered the country in the mid-16th century, and the three States or "Wilayat" of Tripolitania, Cyrenaica and Fezzan (which make up Libya) remained part of their empire with the exception of the virtual autonomy of the Karamanlis. The Karamanlis ruled from 1711 until 1835 mainly in Tripolitania, but had influence in Cyrenaica and Fezzan as well by the mid 18th century. This constituted a first glimpse in recent history of the united and independent Libya that was to re-emerge two centuries later. Reunification came about through the unlikely route of an invasion (Italo-Turkish War, 1911–1912) and occupation starting from 1911 when Italy simultaneously turned the three regions into colonies.

Italian Colony

From 1912 to 1927, the territory of Libya was known as Italian North Africa. From 1927 to 1934, the territory was split into two colonies, Italian Cyrenaica and Italian Tripolitania, run by Italian governors. During the Italian colonial period, between 20% and 50% of the Libyan population died in the struggle for independence, and mainly in prison camps. Some 150,000 Italians settled in Libya, constituting roughly one-fifth of the total population.

In 1934, Italy adopted the name "Libya" (used by the Greeks for all of North Africa, except Egypt) as the official name of the colony (made up of the three provinces of Cyrenaica, Tripolitania and Fezzan). King Idris I, Emir of Cyrenaica, led Libyan resistance to Italian occupation between the two world wars. Between 1928 and 1932 the Italian military "killed half the Bedouin population (directly or through starvation in camps)." From 1943 to 1951, Tripolitania and Cyrenaica were under British administration, while the French controlled Fezzan. In 1944, Idris returned from exile in Cairo but declined to resume permanent residence in Cyrenaica until the removal of some aspects of foreign control in 1947. Under the terms of the 1947 peace treaty with the Allies, Italy relinquished all claims to Libya.

United Kingdom of Libya

On November 21, 1949, the UN General Assembly passed a resolution stating that Libya should become independent before January 1, 1952. Idris represented Libya in the subsequent UN negotiations. On December 24, 1951, Libya declared its independence as the United Kingdom of Libya, a constitutional and hereditary monarchy under King Idris.

The discovery of significant oil reserves in 1959 and the subsequent income from petroleum sales enabled one of the world's poorest nations to establish an extremely wealthy state. Although oil drastically improved the Libyan government's finances, popular resentment began to build over the increased concentration of the nation's wealth in the hands of King Idris and the national elite. This discontent continued to mount with the rise of Nasserism and Arab nationalism throughout North Africa and the Middle East.

Modern Libya

Revolution of 1969

On September 1, 1969, a small group of military officers led by then 27-year-old army officer Muammar al-Gaddafi staged a *coup d'etat* against King Idris, launching the Libyan Revolution. At the time, Idris was in Turkey for medical treatment. His nephew, Crown Prince Sayyid Hasan ar-Rida al-Mahdi as-Sanussi, became King. It was clear that the revolutionary officers who had announced the deposition of Sultan Idris did not want to appoint him over the instruments of state as King. Gaddafi was at the time only a captain and his co-conspirators were all junior officers. Nevertheless the small group seized Libyan military headquarters (due to the sympathies of the stationed men) and the radio broadcasting station with 48 rounds of revolver ammunition. Before the end of September 1, Sayyid Hasan ar-Rida had been formally deposed by the revolutionary army officers and put under house arrest. Meanwhile, revolutionary officers abolished the monarchy, and proclaimed the new Libyan Arab Republic. Gaddafi was, and is to this day, referred to as the "Brother Leader and Guide of the Revolution" in government statements and the official press.

Politics

Libya is considered a de-facto dictatorship by several international definitions and analogies, and is ruled by Colonel Muammar Al-Gaddafi. In theory, there are two branches of government in Libya. The "revolutionary sector" comprises Revolutionary Leader Gaddafi, the Revolutionary Committees and the remaining members of the 12-person Revolutionary Command Council, which was established in 1969. The historical revolutionary leadership is not elected and cannot be voted out of office; they are in power by virtue of their involvement in the revolution.

Constituting the legislative branch of government, this sector comprises Local People's Congresses in each of the 1,500 urban wards, 32 Shabiyat People's Congresses for the regions, and the National General People's Congress. These legislative bodies are represented by corresponding executive bodies (Local People's Committees, Shabiyat People's Committees and the National General People's Committee/Cabinet).

Every four years, the membership of the Local People's Congresses elects their own leaders and the secretaries for the People's Committees, sometimes after many debates and a critical vote. The leadership of the Local People's Congress represents the local congress at the People's Congress of the next level. The members of the National General People's Congress elect the members of the National General People's Committee (the Cabinet) at their annual meeting.

The government controls both state-run and semi-autonomous media. In cases involving a violation of "certain taboos", the private press, like *The Tripoli Post*, has been censored, although articles that are critical of policies have been requested and intentionally published by the revolutionary leadership itself as a means of initiating reforms.

Political parties were banned by the 1972 Prohibition of Party Politics Act Number 71. According to the Association Act of 1971, the establishment of non-governmental organisations (NGOs) is allowed. However, because they are required to conform to the goals of the revolution, their numbers are small in comparison with those in neighbouring countries. Trade unions do not exist, but numerous professional associations are integrated into the state structure as a third pillar, along with the People's Congresses and Committees. These associations do not have the right to strike. Professional associations send delegates to the General People's Congress, where they have a representative mandate.

Foreign Relations

Libya's foreign policies have undergone much fluctuation and change since the state was proclaimed on December 24, 1951. As a Kingdom, Libya maintained a definitively pro-Western stance, yet was recognized as belonging to the conservative traditionalist bloc in the League of Arab States (the present-day Arab League), of which it became a member in 1953. The government was in close alliance with Britain and the United States; both countries maintained military base rights in Libya. Libya also forged close ties with France, Italy, Greece, and established full diplomatic relations with the Soviet Union in 1955.

Although the government supported Arab causes, including the Moroccan and Algerian independence movements, it took little active part in the Arab-Israeli dispute or the tumultuous inter-Arab politics of the 1950s and early 1960s. The Kingdom was noted for its close association with the West, while it steered an essentially conservative course at home.

After the 1969 coup, Gaddafi closed American and British bases and partially nationalized foreign oil and commercial interests in Libya. He also played a key role in promoting oil embargoes as a political weapon for challenging the West, hoping that an oil price rise and embargo in 1973 would persuade the West, especially the United States, to end support for Israel. Gaddafi rejected both Eastern (Soviet) communism and Western (United States) capitalism and claimed he was charting a middle course for his government. In the 1980s, Libya increasingly distanced itself from the United States, based on the principle of non-alignment and the adoption of a middle path between capitalism and communism referred to as "the Third Theory". The animosity was deepened due to Gaddafi's support for groups like the Palestine Liberation Organization, which were considered terrorist by the USA, and his flirtation with the Soviet Union, which at the time represented the sole challenger to the US. Secretary of State Alexander Haig considered Libya as "a Soviet satellite" and a "Soviet-run terrorist training network". When Libya intervened in Chad in 1980 it was perceived by the American authorities as the Soviet Union's attempt to spread control in Africa. In addition to this, Gaddafi's opposition to Israel, a United States ally and considered by them to be the only democratic state in the region, were enough reasons to have Libya considered an American enemy. Consequently, the Reagan administration began its campaign of assisting Libya's neighbours militarily to be able to respond to any Libyan attempt to invade them. Tunisia was given some fifty-four M60 tanks plus $15 million in military credits, while other countries like Egypt and Sudan were given an increase in military credits and training with a full-fledged promise of support in face of Libyan threats. These strategies aimed at isolating Libya and pressure it to reconsider its policies towards the US.

The first confrontation with the United States was when Gaddafi had declared two hundred miles of the Gulf of Sidra to be restricted of any international usage; having defied such declaration Libyan air force fired a missile at a US Boeing EC-135 flight. The attack did not cause any damages to the aircraft, and Jimmy Carter, the U.S. President at the time, did not respond militarily. Allegedly, Gaddafi had secretly ordered the burning down of the US embassy in Tripoli as his fight against the United States. In response U.S. President Ronald Reagan had the "Libyan People's Bureau" closed, and oil imports banned from North African States. Reagan also contested the restricted area defined by Gaddafi based on a 1958 convention that stated that countries were allowed to claim twenty four miles of width from their coasts. On August 19, 1981 the navy was sent close to Libya's coast which resulted in a confrontation where two of the SU-22 fighters supplied to Libya by the Soviet Union were shot down. Following this, Libya was implicated in committing mass acts of state-sponsored terrorism. When CIA allegedly intercepted two messages implying Libyan complicity in the Berlin discotheque terrorist bombing that killed two American servicemen, the United States found this a good enough reason to launch an aerial bombing attack against targets near Tripoli and Benghazi in April 1986. The attack, Operation El Dorado Canyon, was not sanctioned by France and Spain, who refused to allow US F-111 bombers to fly over their territories, and resulted in death of several civilians, including Gaddafi's two-year old adopted daughter.

In 1991, two Libyan intelligence agents were indicted by federal prosecutors in the United States and the United Kingdom for their involvement in the December 1988 bombing of Pan Am flight 103. Six other Libyans were put on trial in absentia for the 1989 bombing of UTA Flight 772 over Chad and Niger. The UN Security Council demanded that Libya surrender the suspects, cooperate with the Pan Am 103 and UTA 772 investigations, pay compensation to the victims' families, and cease all support for terrorism. Libya's refusal to comply led to the approval of Security Council Resolution 748 on March 31, 1992, imposing sanctions on the state designed to bring about Libyan compliance. Continued Libyan defiance led to further sanctions by the UN against Libya in November 1993.

In 1999, less than a decade after the sanctions were put in place, Libya began to make dramatic policy changes in regard to the Western world, including turning over the Lockerbie suspects for trial. This diplomatic breakthrough followed years of negotiation, including a visit by UN Secretary General Kofi Annan to Libya in December 1998, and personal appeals by Nelson Mandela. Eventually UK Foreign Secretary Robin Cook persuaded the Americans to accept a trial of the suspects in the Netherlands under Scottish law, with the UN Security Council agreeing to suspend sanctions as soon as the suspects arrived in the Netherlands for trial.

In response to 9/11 attacks Gaddafi condemned the attacks as an act of terrorism and urged Libyans to donate blood for the US victims. However, the United States were still not willing to remove the sanctions of Libya yet. After the invasion of Iraq based on allegations that it had WMD programs violating non-proliferation treaty, and the fall of Saddam Hussein in

2003, the Libyan government announced its decision to abandon its weapons of mass destruction programmes and pay almost 3 billion US dollars in compensation to the families of Pan Am flight 103 as well as UTA Flight 772. According to some sources Gaddafi had privately phoned Italian Premier Silvio Berlusconi expressing his fear that his regime would meet the same fate if he did not take such steps. The decision was welcomed by many western nations and was seen as an important step for Libya toward rejoining the international community. Since 2003 the country has made efforts to normalize its ties with the European Union and the United States and has even coined the catchphrase, 'The Libya Model', an example intended to show the world what can be achieved through negotiation rather than force when there is goodwill on both sides. By 2004 Bush had lifted the economic sanctions on Libya and official relations resumed between Libya and the United States. Libya then opened a Liaison office in Washington, DC and the United States opened an office in Tripoli. In January 2004, Congressman Tom Lantos led the first official Congressional delegation visit to Libya.

An event considered pivotal by many in Libyan-Western relations is the HIV trials (1999–2007) of five Bulgarian nurses and a Palestinian doctor. Their release is seen as marking a new stage in Libyan-Western relations.

On May 15, 2006 the United States State Department announced it would fully restore diplomatic relations with Libya if it dismantled its weapons programmes. The State Department also removed Libya from their state sponsored terrorism list which it had been on for 27 years. This move has also been attributed to the pressures of oil companies lobbying the Congress. In addition to that the fall of the Soviet power, the prominent role that Libya plays in the African Continent, and the assistance it could provide to the US in its war on terror are among the other considerations that were factored in. In August 2008 a motion was introduced in the 110th Congress known as S 3370 or the "Libyan Claims Resolution Act" to exempt Libya from the infamous section 1083 clause of the National Defence Authorization Act. The motion passed both the House of Representatives and the Senate by unanimous consent, and is signed into law by President George W. Bush on 4 August. After Libya paid a final portion of $1.8 billion global settlement fund for American victims it became formally exempted from section 1083. Following that Libyan families received $300 million for casualties suffered due to the 1986 airstrikes led by the United States. In November the same year, the United States Senate confirmed Gene A. Cretz as the first US Ambassador to Libya in over 35 years. The final step in the process of rebuilding the relations between the two countries came in January 2009 when Ali Suleiman Aujali presented his letters of credentials to President George W. Bush as Ambassador Extraordinaire and Plenipotentiary of Libya to the United States of America, and Gene A. Cretz presents his letter of credentials before the General People's Congress; currently both are serving as Ambassadors to their respective countries.

On October 16, 2007, Libya was voted to serve on the United Nations Security Council for two years starting January 2008.

In February 2009, Gaddafi was selected to be chairman of the African Union for one year.

As of October 25, 2009, Canadian visa requests are being denied and Canadian travellers have been told they're not welcome in Libya, in an apparent reprisal for Canada's near tongue-lashing of Muammar Gaddafi. Meanwhile, Libya is still detaining two Swiss businessmen. Libyan-Swiss relations strongly suffered after the arrest of Hannibal Gadhafi for beating up his domestic servants in Geneva in 2008. This was an embarrassment for dictator Ghadafi who removed all his money held in Swiss banks and wanted the UN to vote to abolish Switzerland as a Soverign nation

Cooperation with Italy

On August 30, 2008, Gaddafi and Italian Prime Minister Silvio Berlusconi signed a historic cooperation treaty in Benghazi. Under its terms, Italy will pay $5 billion to Libya as compensation for its former military occupation. In exchange, Libya will take measures to combat illegal immigration coming from its shores and boost investments in Italian companies.

The treaty was ratified by Italy on February 6, 2009, and by Libya on March 2, during a visit to Tripoli by Berlusconi. In June Gaddafi made his first visit to Rome, where he met Prime Minister Berlusconi, President Giorgio Napolitano, Senate President Renato Schifani, and Chamber President Gianfranco Fini, among others. The Democratic Party and Italy of Values opposed the visit, and many protests were staged throughout Italy by human rights organizations and the Radical Party. Gaddafi also took part in the G8 summit in L'Aquila in July 2009 as Chairman of the African Union.

Human Rights

According to the U.S. Department of State's annual human rights report for 2007, Libya's authoritarian regime continued to have a poor record in the area of human rights. Some of the numerous and serious abuses on the part of the government include poor prison conditions, arbitrary arrest and prisoners held incommunicado, and political prisoners held for many years without charge or trial. The judiciary is controlled by the government, and there is no right to a fair public trial. Libyans do not have the right to change their government. Freedom of speech, press, assembly, association, and religion are restricted. Independent human rights organizations are prohibited. Ethnic and tribal minorities suffer discrimination, and the state continues to restrict the labour rights of foreign jobs.

In 2005 Freedom House rated political rights in Libya as "7" (1 representing the most free and 7 the least free rating), civil liberties as "7" and gave it the freedom rating of "Not Free".

Administrative Divisions

Historically the area of Libya was considered three provinces (or states), Tripolitania in the northwest, Barka (Cyrenaica) in the east, and Fezzan in the southwest. It was the conquest by Italy in the Italo-Turkish War that united them in a single political unit. Under the Italians Libya, in 1934, was divided into four provinces and one territory (in the south): Tripoli, Misurata, Benghazi, Al Bayda, and the Territory of the Libyan Sahara. After independence, Libya was divided into three governorates (*muhafazat*) and then in 1963 into ten governorates. The governorates were legally abolished in February 1975, and nine "control bureaus" were set up to deal directly with the nine areas, respectively: education, health, housing, social services, labour, agricultural services, communications, financial services, and economy, each under their own ministry. However, the courts and some other agencies continued to operate as if the governorate structure were still in place. In 1983 Libya was split into forty-six districts (*baladiyat*), then in 1987 into twenty-five. In 1995, Libya was divided into thirteen districts (*shabiyah*), in 1998 into twenty-six districts, and in 2001 into thirty-two districts.

Geography

Libya extends over 1,759,540 square kilometres (679,362 sq mi), making it the 17th largest nation in the world by size. Libya is somewhat smaller than Indonesia, and roughly the size of the US state of Alaska. It is bound to the north by the Mediterranean Sea, the west by Tunisia and Algeria, the southwest by Niger, the south by Chad and Sudan and to the east by Egypt. At 1,770 kilometres (1,100 mi), Libya's coastline is the longest of any African country bordering the Mediterranean. The portion of the Mediterranean Sea north of Libya is often called the Libyan Sea. The climate is mostly dry and desertlike in nature. However, the northern regions enjoy a milder Mediterranean climate.

Natural hazards come in the form of hot, dry, dust-laden sirocco (known in Libya as the *gibli*). This is a southern wind blowing from one to four days in spring and autumn. There are also dust storms and sandstorms. Oases can also be found scattered throughout Libya, the most important of which are Ghadames and Kufra.

Libyan Desert

The Libyan Desert, which covers much of Libya, is one of the most arid places on earth. In places, decades may pass without rain, and even in the highlands rainfall seldom happens, once every 5–10 years. At Uweinat, as of 2006 the last recorded rainfall was in September 1998. There is a large depression, the Qattara Depression, just to the south of the northernmost scarp, with Siwa oasis at its western extremity. The depression continues in a shallower form west, to the oases of Jaghbub and Jalo.

Likewise, the temperature in the Libyan desert can be extreme; in 1922, the town of Al 'Aziziyah, which is located Southwest of Tripoli, recorded an air temperature of 57.8 °C (136.0 °F), generally accepted as the highest recorded naturally occurring air temperature reached on Earth.

There are a few scattered uninhabited small oases, usually linked to the major depressions, where water can be found by digging to a few feet in depth. In the west there is a widely dispersed group of oases in unconnected shallow depressions, the Kufra group, consisting of Tazerbo, Rebianae and Kufra. Aside from the scarps, the general flatness is only interrupted by a series of plateaus and massifs near the centre of the Libyan Desert, around the convergence of the Egyptian-Sudanese-Libyan borders.

Slightly further to the south are the massifs of Arkenu, Uweinat and Kissu. These granite mountains are ancient, having formed long before the sandstones surrounding them. Arkenu and Western Uweinat are ring complexes very similar to those in the Air Mountains. Eastern Uweinat (the highest point in the Libyan Desert) is a raised sandstone plateau adjacent to the granite part further west. The plain to the north of Uweinat is dotted with eroded volcanic features. With the discovery of oil in the 1950s also came the discovery of a massive aquifer underneath much of the country. The water in this aquifer pre-dates the last ice ages and the Sahara desert itself. The country is also home to the Arkenu craters, double impact craters found in the desert.

Economy

The Libyan economy depends primarily upon revenues from the oil sector, which constitute practically all export earnings and about one-quarter of gross domestic product (GDP). The World Bank defines Libya as an 'Upper Middle Income Economy', along with only seven other African countries. In the early 1980s, Libya was one of the wealthiest countries in the world; its GNP per capita was higher than that of countries such as Italy, Singapore, South Korea, Spain and New Zealand.

Today, high oil revenues and a small population give Libya one of the highest GDPs per person in Africa and have allowed the Libyan state to provide an extensive level of social security, particularly in the fields of housing and education. Many problems still beset Libya's economy however; unemployment is the highest in the region at 21% according to the latest census figures.

Compared to its neighbours, Libya enjoys a low level of both absolute and relative poverty. Libyan officials in the past six years have carried out economic reforms as part of a broader campaign to reintegrate the country into the global capitalist economy. This effort picked up steam after UN sanctions were lifted in September 2003, and as Libya announced in December 2003 that it would abandon programmes to build weapons of mass destruction.

Libya has begun some market-oriented reforms. Initial steps have included applying for membership of the World Trade Organization, reducing subsidies, and announcing plans for privatisation. Authorities have privatised more than 100 government owned companies since 2003 in industries including oil refining, tourism and real estate, of which 29 are 100% foreign owned. The non-oil manufacturing and construction sectors, which account for about 20% of GDP, have expanded from processing mostly agricultural products to include the production of petrochemicals, iron, steel and aluminium.

Climatic conditions and poor soils severely limit agricultural output, and Libya imports about 75% of its food. Water is also a problem, with some 28% of the population not having access to safe drinking water in 2000. The Great Manmade River project is tapping into vast underground aquifers of fresh water discovered during the quest for oil, and is intended to improve the country's agricultural output.

Under the previous Prime Minister, Shukri Ghanem, and current prime minister Baghdadi Mahmudi, Libya is undergoing a business boom. Many government-run industries are being privatised. Many international oil companies have returned to the country, including oil giants Shell and Exxon Mobil.

Tourism is on the rise, bringing increased demand for hotel accommodation and for capacity at airports such as Tripoli

International. A multi-million dollar renovation of Libyan airports has recently been approved by the government to help meet such demands. At present 130,000 people visit the country annually; the Libyan government hopes to increase this figure to 10,000,000 tourists. Saif al-Islam al-Gaddafi, the second-eldest son of Muammar al-Gaddafi, is involved in a green development project called the Green Mountain Sustainable Development Area, which seeks to bring tourism to Cyrene and to preserve Greek ruins in the area.

Demographics

Libya has a small population residing in a large land area. Population density is about 50 persons per km² (80/sq. mi.) in the two northern regions of Tripolitania and Cyrenaica, but falls to less than one person per km² (1.6/sq. mi.) elsewhere. Ninety percent of the people live in less than 10% of the area, primarily along the coast. About 88% of the population is urban, mostly concentrated in the two largest cities, Tripoli and Benghazi. 50% of the population is estimated to be under age 15.

Native Libyans are primarily Arabs (mainly tribal desert Arabs "Bedouins"), Berbers and arabized Berbers, Tuareg. Small Hausa, and Tebu tribal groups in southern Libya are nomadic or seminomadic. Among foreign residents, the largest groups are citizens of other African nations, including North Africans (primarily Egyptians), and Sub-Saharan Africans. Libya is home to a large illegal population which numbers more than one million. Libya has a small Italian minority. Previously, there was a visible presence of Italian settlers, but many left after independence in 1947 and many more left after the rule of Muammar al-Gaddafi in 1970.

The main language spoken in Libya is Arabic by 80% of the Libyans, and which is also the official language; the Tamazight spoken by 20% (i.e. Berber and Tuareg languages), which do not have official status, are spoken by Libyan Berbers and Tuaregs in the south beside Arabic language. Berber speakers live above all in the Jebel Nafusa region (Tripolitania), the town of Zuwarah on the coast, and the city-oases of Ghadames, Ghat and Awjila. In addition, Tuaregs speak Tamahaq, the only known Northern Tamasheq language, also Toubou language is spoken by Toubou in some pockets in Qatroun village and Koffra city. Italian and English are sometimes spoken in the big cities, although Italian speakers are mainly among the older generation.

Family life is important for Libyan families, the majority of which live in apartment blocks and other independent housing units, with precise modes of housing depending on their income and wealth. Although the Libyan Arabs traditionally lived nomadic lifestyles in tents, they have now settled in various towns and cities. Because of this, their old ways of life are gradually fading out. An unknown small number of Libyans still live in the desert as their families have done for centuries. Most of the population has occupations in industry and services, and a small percentage is in agriculture.

According to the *World Refugee Survey 2008*, published by the U.S. Committee for Refugees and Immigrants, Libya hosted a population of refugees and asylum seekers numbering approximately 16,000 in 2007. Of this group, approximately 9,000 persons were from the Former Palestine, 3,200 from Sudan, 2,500 from Somalia and 1,100 from Iraq. Libya reportedly deported thousands of illegal entrants in 2007 without giving them the opportunity to apply for asylum. Refugees faced discrimination from Libyan officials when moving in the country and seeking employment.

Education

Libya's population includes 1.7 million students, over 270,000 of whom study at the tertiary level. Education in Libya is free for all citizens, and compulsory up until secondary level. The literacy rate is the highest in North Africa; over 82% of the population can read and write.

After Libya's independence in 1951, its first university, the University of Libya, was established in Benghazi. In academic year 1975/76 the number of university students was estimated to be 13,418. As of 2004, this number has increased to more than 200,000, with an extra 70,000 enrolled in the higher technical and vocational sector. The rapid increase in the number of students in the higher education sector has been mirrored by an increase in the number of institutions of higher education. Since 1975 the number of universities has grown from two to nine and after their introduction in 1980, the number of higher technical and vocational institutes currently stands at 84 (with 12 public universities). Libya's higher education is financed by the public budget. In 1998 the budget allocated for education represented 38.2% of the national budget.

The main universities in Libya are:

- Al Fateh University (Tripoli)
- Garyounis University (Benghazi).

Religion

Religion	*Percent*
Islam	97%
Christianity	2%
All Others	1%

By far the predominant religion in Libya is Islam with 97% of the population associating with the faith. The vast majority of Libyan Muslims adhere to Sunni Islam, which provides both a spiritual guide for individuals and a keystone for government policy, but a minority (between 5 and 10%) adhere to Ibadism (a branch of Kharijism), above all in the Jebel Nefusa and the town of Zuwarah, west of Tripoli.

Before the 1930s, the Senussi Movement was the primary Islamic movement in Libya. This was a religious revival adapted to desert life. Its *zawaaya* (lodges) were found in Tripolitania and Fezzan, but Senussi influence was strongest in Cyrenaica. Rescuing the region from unrest and anarchy, the Senussi movement gave the Cyrenaican tribal people a religious attachment and feelings of unity and purpose.

This Islamic movement, which was eventually destroyed by both Italian invasion and later the Gaddafi government, was very conservative and somewhat different from the Islam that exists in Libya today. Gaddafi asserts that he is a devout Muslim, and his government is taking a role in supporting Islamic institutions and in worldwide proselytizing on behalf of Islam. A Libyan form of Sufism is also common in parts of the country.

Other than the overwhelming majority of Sunni Muslims, there are also small foreign communities of Christians. Coptic Orthodox Christianity, which is the Christian Church of Egypt, is the largest and most historical Christian denomination in Libya. There are over 60,000 Egyptian Copts in Libya, as they comprise over 1% of the population. There are an estimated 40,000 Roman Catholics in Libya who are served by two Bishops, one in Tripoli (serving the Italian community) and one in Benghazi (serving the Maltese community). There is also a small Anglican community, made up mostly of African immigrant workers in Tripoli; it is part of the Anglican Diocese of Egypt.

Libya was until recent times the home of one of the oldest Jewish communities in the world, dating back to at least 300 BC. In the 1st century, the Jewish historian for the Roman empire, Joseph Flavius, noted that 500,000 Jews lived in Libya.

In 1942, under Fascist Italian orders, the Libyan Muslims instituted several forced labour camps south of Tripoli for the Jews, including Giado (about 3,000 Jews) and Gharyan, Jeren, and Tigrinna. In Giado some 500 Jews died of weakness, hunger, and disease. Summer-Fall of 1942, Jews who were not in the concentration camps were heavily restricted in their economic activity. All Jewish males, 18 to 45 years, were drafted for forced labour. In August 1942, the concentration camp Sidi Azaz interned Jews from the Tripolitania region. In October 350 Jews were deported to the Tobruk area.

Libya was liberated from the Italians on January 23, 1943. The Muslims of Libya responded with a three-day pogrom (November 5–7, 1945) against the Jews. More than 140 Jews were murdered, hundreds more were wounded. This series of pogroms beginning in November 1945 lasted for almost three years, drastically reducing Libya's Jewish population. In 1948, about 38,000 Jews remained in the country. Upon Libya's independence in 1951, most of the Jewish community emigrated.

Culture

Libya is culturally similar to its neighbouring Maghrebian states. Libyans consider themselves very much a part of a wider Arab community. The Libyan state tends to strengthen this feeling by considering Arabic as the only official language, and forbidding the teaching and even the use of the Berber language. Libyan Arabs have a heritage in the traditions of the nomadic Bedouin and associate themselves with a particular Bedouin tribe.

As with some other countries in the Arab world, Libya boasts few theatres or art galleries. For many years there have been no public theatres, and only a few cinemas showing foreign films. The tradition of folk culture is still alive and well, with troupes performing music and dance at frequent festivals, both in Libya and abroad.

The main output of Libyan television is devoted to showing various styles of traditional Libyan music. Tuareg music and dance are popular in Ghadames and the south. Libyan television programmes are mostly in Arabic with a 30-minute news broadcast each evening in English and French.

The government maintains strict control over all media outlets. A new analysis by the Committee to Protect Journalists has found Libya's media the most tightly controlled in the Arab world. To combat this, the government plans to introduce private media, an initiative intended to bring the country's media in from the cold.

Many Libyans frequent the country's beach and they also visit Libya's beautifully preserved archaeological sites—especially Leptis Magna, which is widely considered to be one of the best preserved Roman archaeological sites in the world.

The nation's capital, Tripoli, boasts many good museums and archives; these include the Government Library, the Ethnographic Museum, the Archaeological Museum, the National Archives, the Epigraphy Museum and the Islamic Museum. The Jamahiriya Museum, built in consultation with UNESCO, may be the country's most famous. It houses one of the finest collections of classical art in the Mediterranean.

Contemporary Travel

The most common form of public transport between cities is the bus, but many people do travel by automobile. There are no railway services in Libya.

Libyan Cuisine

Libyan cuisine is generally simple, and is very similar to Sahara cuisine. In many undeveloped areas and small towns, restaurants may be nonexistent, and food stores may be the only source to obtain food products. Some common Libyan foods include couscous, bazeen, which is a type of unsweetened cake, and shurba, which is soup. Libyan restaurants may serve international cuisine, or may serve simpler fare such as lamb, chicken, vegetable stew, potatoes and macaroni. Alcohol consumption is illegal in the entire country, and this law is enforced in Libya.

11

The Gabon Republic

Gabon is a country in west central Africa sharing borders with the Gulf of Guinea to the west, Equatorial Guinea to the northwest, and Cameroon to the north, with the Republic of the Congo curving around the east and south. It covers a land area of nearly 270,000 km² and has an estimated population of 1,500,000. Its capital and largest city is Libreville.

Since its independence from France on August 17, 1960, Gabon has been ruled by three presidents. In the early 1990s, Gabon introduced a multi-party system and a new democratic constitution that allowed for a more transparent electoral process and reformed many governmental institutions.

The small population density together with abundant natural resources and foreign private investment have helped make Gabon one of the most prosperous countries in the region, with the highest HDI in Sub-Saharan Africa.

History

The earliest inhabitants of the area were Pygmy peoples. They were largely replaced and absorbed by Bantu tribes as they migrated.

In the 15th century, the first Europeans arrived. The nation's present name originates from "Gabao", Portuguese for "cloak", which is roughly the shape of the estuary of the Komo River by Libreville. French explorer Pierre Savorgnan de Brazza led his first mission to the Gabon-Congo area in 1875. He founded the town of Franceville, and was later colonial governor. Several

Bantu groups lived in the area that is now Gabon when France officially occupied it in 1885.

In 1910, Gabon became one of the four territories of French Equatorial Africa, a federation that survived until 1959. These territories became independent on August 17, 1960. The first president of Gabon, elected in 1961, was Léon M'ba, with Omar Bongo Ondimba as his vice president. French interests were decisive in selecting the future leadership in Gabon after Independence; French logging interests poured funds into the successful election campaign of M'ba, an 'evolue' from the coastal region.

After M'ba's accession to power, the press was suppressed, political demonstrations banned, freedom of expression curtailed, other political parties gradually excluded from power and the Constitution changed along French lines to vest power in the Presidency, a post that M'ba assumed himself. However, when M'ba dissolved the National Assembly in January 1964 to institute one-party rule, an army coup sought to oust him from power and restore parliamentary democracy. The extent to which M'ba's dictatorial regime was synonymous with "French Interests" then became blatantly apparent when French paratroopers flew in within 24 hours to restore M'ba to power.

After a few days of fighting, the coup was over and the opposition imprisoned, despite widespread protests and riots. The French government was unperturbed by international condemnation of the intervention; and paratroops still remain in the Camp de Gaulle on the outskirts of Gabon's capital. When M'Ba died in 1967, Bongo replaced him as president, and continued to be the head of state until his death in 2009, winning each contested election with a substantial majority.

Government

In March 1991, a new constitution was enacted. Among its provisions are a bill of rights, the creation of a body to guarantee those rights (National Council of Democracy) and a governmental advisory board which deals with economic and social issues. Multi-party legislative elections were held in 1990–91 even though opposition parties had not yet been formally declared legal.

President El Hadj Omar Bongo Ondimba, in power from 1967 until his death in June 2009, was re-elected to his third consecutive seven-year term on November 27, 2005. According to figures provided by Gabon's Interior Ministry, he received a 79.1% majority of votes. Succeeding Bongo as interim president upon his death was Rose Francine Rogombe, President of the Senate of Gabon.

The voting age in Gabon is 21 years of age. In 2003, the President amended the Constitution of Gabon to remove any restrictions on the number of terms a president is allowed to serve. The president retains strong powers, such as authority to dissolve the National Assembly, declare a state of siege, delay legislation, conduct referendums, and appoint or dismiss the prime minister as well as cabinet members. In provisional results, the ruling Gabonese Democratic Party (PDG) won 84 out of 120 parliamentary seats.

On September 3, 2009, the son of Omar Bongo, Ali Ben Bongo, was elected president. As with previous Gabonese elections, the opposition parties have contested the results. There were calls for a boycott and accusations of electoral fraud and bribery. The announcement of the result sparked looting and the torching of the French consulate in Port-Gentil. However, several international observers including the Economic Community of Central African States have reported that the election "met international standards" for democratic voting, and urged the people of Gabon to accept the result.

Gabon has a small, professional military of about 5,000 personnel, divided into army, navy, air force, gendarmerie, and national police. Gabonese forces are oriented to the defence of the country and have not been trained for an offensive role. A 1,800-member guard provides security for the president.

In September, 2007, René Ndemezo'o Obiang, the government's spokesperson, announced that Gabon's cabinet council had decided to formally abolish the death penalty, which had not been applied in the country in over a decade.

Gabon placed 21st out of 53 African countries in the 2009 Ibrahim Index of African Governance. The Ibrahim Index is a comprehensive measure of African governance, based on a number of different variables which reflect the success with which governments deliver essential political goods to its citizens.

Provinces and Departments

Gabon is divided into nine provinces and further divided into 37 departments.

The provinces are:

1. Estuaire
2. Haut-Ogooue
3. Moyen-Ogooue
4. Ngounie
5. Nyanga
6. Ogooue-Ivindo
7. Ogooue-Lolo
8. Ogooue-Maritime
9. Woleu-Ntem.

Geography

Gabon is located on the Atlantic coast of central Africa. Located on the equator, Gabon has an equatorial climate with an extensive system of rainforests covering 85% of the country. There are three distinct regions: the coastal plains (ranging between 20 to 300 km from the ocean's shore), the mountains (the Cristal Mountains to the northeast of Libreville, the Chaillu Massif in the centre, culminating at 1575 m with Mont Iboundji), and the savanna in the east. The coastal plains form a large section of the World Wildlife Fund's Atlantic Equatorial coastal forests ecoregion and contain patches of Central African mangroves especially on the Muni River estuary on the border with Equatorial Guinea.

Gabon's largest river is the Ogooue which is 1200 km long. Gabon has three karst areas where there are hundreds of caves located in the dolomite and limestone rocks. Some of the caves include Grotte du Lastoursville, Grotte du Lebamba, Grotte du Bongolo, and Grotte du Kessipougou. Many caves have not been explored yet. A National Geographic Expedition visited the caves in the summer of 2008 to document them.

Gabon is also noted for efforts to preserve the natural environment. In 2002, President Omar Bongo Ondimba put Gabon firmly on the map as an important future ecotourism destination by designating more than 11% of the nation's territory to be part of its national park system (with 13 parks in total), one of the largest proportions of nature parkland in the world. Natural resources include: petroleum, magnesium, iron, gold, uranium, and forests.

Economy

Gabon is more prosperous than most nearby countries, with a per capita income of four times the average for Sub-Saharan Africa. This is in large part due to offshore oil production. Critics note that the income was not invested in modernizing or diversifying the economy and Gabon remains heavily reliant on its natural resources. Gabon was a full member of OPEC from 1975 to 1995. It is an exporter of manganese, iron, and wood. Uranium mines near Franceville were shut down in 2001 with the arrival of new competition on the global market and there is work in progress to re-open them. Planned efforts to exploit rich iron deposits north-east of Makokou are expected to begin in 2012.

During the 1990s, devaluation of the CFA franc left Gabon struggling to pay its overseas debt; France and the IMF have provided further loans and aid in exchange for the implementation of changes to the economy. Gabon's principal trading partners are the United States, China, and Russia for exports while importing mainly from France.

On December 5, 2007 JP Morgan acted as Joint-Bookrunner on the Gabonese Republic's (BB-/BB-) debut international US$1 billion 10-year bond issue.

Gabon is a member of the Organization for the Harmonization of Business Law in Africa (OHADA).

Tourism

Straddling the equator, this rainforest nation is situated on West Africa's Atlantic coast. Animals are abundant – Gabon is the continent's fifth least densely populated country – and is 85 per cent covered in tropical forest. This, along with savannahs, mangroves, lagoons and beaches makes it an ideal habitat for varied species of animals and reptiles including 20,000 western lowland gorillas and 60,000 forest elephants – the largest population in Africa – and 700 types of exotic birds.

The National Tourism Strategy, written in consultation with The Wildlife Conservation Society, aims to achieve 100,000 visitors per annum.

Demographics

The population of Gabon is nearly 1.5 million (1,442,334). Almost all Gabonese are of Bantu origin, though Gabon has at least forty ethnic groups with diverse languages and cultures. The Fang are generally thought to be the largest, although recent census data seem to favour the Bandjabi (or Nzebi). Others include the Myene, Bakota, Eshira, Bapounou, and Okande. Ethnic group boundaries are less sharply drawn in Gabon than elsewhere in Africa. French, the official language, may be regarded as a unifying force. It is estimated that 80% of the country's population are able to speak French, and that 30% of Libreville residents are native speakers of the language. Nationally, 32% of the Gabonese people speak the Fang language as a mother tongue. More than 10,000 French people live in Gabon, and France is the predominant foreign cultural and commercial influence.

Historical and environmental factors caused Gabon's population to decline between 1900 and 1940. It has one of the lowest population densities of any country in Africa, and labour shortages form a major obstacle to development and a draw for foreign workers. Most inhabitants are Christians, with estimates of the Christian population ranging from 55 to 77%, mostly members of the Roman Catholic Church. Other religious groups include animists, Muslims, and practitioners of indigenous African religions. Gabon's literacy rate is 63.2%.

Culture

Gabonese music is lesser-known in comparison with regional giants like the Democratic Republic of the Congo and Cameroon. The country boasts an array of folk styles, as well as pop stars like Patience Dabany and Annie Flore Batchiellilys, a Gabonese singer and renowned live performer. Also known are guitarists like Georges Oyendze, La Rose Mbadou and Sylvain Avara, and the singer Oliver N'Goma. Imported rock and hip hop from the US and UK are popular in Gabon, as are rumba, makossa and soukous. Gabonese folk instruments include the obala, the ngombi, balafon and traditional drums.

A country with a primarily oral tradition up until the spread of literacy in the 21st century, Gabon is rich in folklore and mythology. "Raconteurs" are currently working to keep traditions alive such as the mvett among the Fangs and the ingwala among the Nzebis.

Gabon also features internationally celebrated masks, such as the n'goltang (Fang) and the relicary figures of the Kota. Each group has its own set of masks used for various reasons. They are mostly used in traditional ceremonies such as marriage, birth and funerals. Traditionalists mainly work with rare local woods and other precious materials.

The Federation of the Cameroon

The Republic of Cameroon (French: *Republique du Cameroun*) is a country of central and western Africa. It is bordered by Nigeria to the west; Chad to the northeast; the Central African Republic to the east; and Equatorial Guinea, Gabon, and the Republic of the Congo to the south. Cameroon's coastline lies on the Bight of Bonny, part of the Gulf of Guinea and the Atlantic Ocean. The country is called "Africa in miniature" for its geological and cultural diversity. Natural features include beaches, deserts, mountains, rainforests, and savannas. The highest point is Mount Cameroon in the southwest, and the largest cities are Douala, Yaounde, and Garoua. Cameroon is home to over 200 different linguistic groups. The country is well known for its native styles of music, particularly makossa and bikutsi, and for its successful national football team. English and French are the official languages.

Early inhabitants of the territory included the Sao civilisation around Lake Chad and the Baka hunter-gatherers in the southeastern rainforest. Portuguese explorers reached the coast in the 15th century and named the area *Rio dos Camaroes* ("River of Shrimp"), the name from which *Cameroon* derives. Fulani soldiers founded the Adamawa Emirate in the north in the 19th century, and various ethnic groups of the west and northwest established powerful chiefdoms and fondoms. Cameroon became a German colony in 1884.

After World War I, the territory was divided between France and Britain as League of Nations mandates. The Union des Populations du Cameroun political party advocated independence but was outlawed by France in the 1950s. It waged war on French and UPC militants forces until 1971. In 1960, the French administered part of Cameroon became independent as the Republic of Cameroun under President Ahmadou Ahidjo. The southern part of British Cameroons merged with it in 1961 to form the Federal Republic of Cameroon. The country was renamed the United Republic of Cameroon in 1972 and the Republic of Cameroon in 1984.

Compared to other African countries, Cameroon enjoys relatively high political and social stability. This has permitted the development of agriculture, roads, railways, and large petroleum and timber industries. Nevertheless, large numbers of Cameroonians live in poverty as subsistence farmers. Power lies firmly in the hands of the president, Paul Biya, and his Cameroon People's Democratic Movement party. The English speaking territories of Cameroon have grown increasingly alienated from the government, and politicians from those regions have called for greater decentralization and even the secession (e.g.: the Southern Cameroons National Council) of the former British-governed territories.

History

The territory of present day Cameroon was first settled during the Neolithic. The longest continuous inhabitants are groups such as the Baka. The Sao culture arose around Lake Chad c. AD 500 and gave way to the Kanem and its successor state, the Bornu empire. Kingdoms, fondoms, and chiefdoms arose in the west.

Portuguese sailors reached the coast in 1472. They noted an abundance of prawns and crayfish in the Wouri River and named it *Rio dos Camaroes*, Portuguese for "River of Shrimp", and the phrase from which *Cameroon* is derived. Over the following few centuries, European interests regularised trade with the coastal peoples, and Christian missionaries pushed inland. In the early 19th century, Modibo Adama led Fulani soldiers on a jihad in the north against non-Muslim and partially Muslim peoples and established the Adamawa Emirate. Settled peoples who fled the Fulani caused a major redistribution of population.

The German Empire claimed the territory as the colony of Kamerun in 1884 and began a steady push inland. They initiated projects to improve the colony's infrastructure, relying on a harsh system of forced labour. With the defeat of Germany in World War I, Kamerun became a League of Nations mandate territory and was split into French Cameroun and British Cameroons in 1919. France integrated the economy of Cameroun with that of France and improved the infrastructure with capital investments, skilled workers, and continued forced labour.

The British administered their territory from neighbouring Nigeria. Natives complained that this made them a neglected "colony of a colony". Nigerian migrant workers flocked to Southern Cameroons, ending forced labour but angering indigenous peoples. The League of Nations mandates were converted into United Nations Trusteeships in 1946, and the question of independence became a pressing issue in French Cameroun. France outlawed the most radical political party, the Union des Populations du Cameroun (UPC), on 13 July 1955. This prompted a long guerrilla war and the assassination of the party's leader, Ruben Um Nyobe while on a trip in Switzerland. In British Cameroons, the question was whether to reunify with French Cameroun or join Nigeria.

On 1 January 1960, French Cameroun gained independence from France under President Ahmadou Ahidjo. On 1 October 1961, the formerly British Southern Cameroons united with French Cameroun to form the Federal Republic of Cameroon. Ahidjo used the ongoing war with the UPC to concentrate power in the presidency, continuing with this even after the suppression of the UPC in 1971.

His political party, the Cameroon National Union (CNU), became the sole legal political party on 1 September 1966 and in 1972, the federal system of government was abolished in favour of a United Republic of Cameroon, headed from Yaounde. Ahidjo pursued an economic policy of planned liberalism, prioritising cash crops and petroleum exploitation. The government used oil money to create a national cash reserve, pay farmers, and finance major development projects; however, many initiatives failed when Ahidjo appointed unqualified allies to direct them.

Ahidjo stepped down on 4 November 1982 and left power to his constitutional successor, Paul Biya. However, Ahidjo remained in control of the CNU and tried to run the country from behind the scenes until Biya and his allies pressured him into resigning. Biya began his administration by moving toward a more democratic government, but a failed coup d'etat nudged him toward the leadership style of his predecessor.

An economic crisis took effect in the mid-1980s to late 1990s as a result of international economic conditions, drought, falling petroleum prices, and years of corruption, mismanagement, and cronyism. Cameroon turned to foreign aid, cut government spending, and privatised industries. With the reintroduction of multi-party politics in December 1990, the former British Cameroons pressure groups called for greater autonomy, with some (SCNC) advocating complete secession as the Republic of Ambazonia. In February 2008, Cameroon experienced its worst violence in 15 years when a transport union strike in Douala escalated into violent protests in 31 municipal areas.

Politics and Government

The President of Cameroon has broad, unilateral powers to create policy, administer government agencies, command the armed forces, negotiate and ratify treaties, and declare a state of emergency. The president appoints government officials at all levels, from the prime minister (considered the official head of government), to the provincial governors, divisional officers, and urban-council members in large cities. The president is selected by popular vote every seven years. In smaller municipalities, the public elects mayors and councilors.

Corruption is rife at all levels of government. In 1997, Cameroon established anti-corruption bureaus in 29 ministries, but only 25% became operational, and in 2007, Transparency International placed Cameroon at number 138 on a list of 163 countries ranked from least to most corrupt. On 18 January 2006, Biya initiated an anti-corruption drive under the direction of the National Anti-Corruption Observatory.

Cameroon's legal system is largely based on French civil law with common law influences. Although nominally independent, the judiciary falls under the authority of the executive's Ministry of Justice. The president appoints judges at all levels. The judiciary is officially divided into tribunals, the court of appeal, and the supreme court. The National Assembly elects the members of a nine-member High Court of Justice that judges high-ranking members of government in the event they are charged with high treason or harming national security.

Human rights organisations accuse police and military forces of mistreating and even torturing criminal suspects, ethnic minorities, homosexuals, and political activists. Prisons are overcrowded with little access to adequate food and medical facilities, and prisons run by traditional rulers in the north are charged with holding political opponents at the behest of the government. However, since the early 2000s, an increasing number of police and gendarmes have been prosecuted for improper conduct.

The National Assembly makes legislation. The body consists of 180 members who are elected for five-year terms and meet three times per year. Laws are passed on a majority vote. Rarely has the assembly changed or blocked legislation proposed by the president. The 1996 constitution establishes a second house of parliament, the 100-seat Senate, but this body has never been put into practice. The government recognises the authority of traditional chiefs, fons, and lamibe to govern at the local level and to resolve disputes as long as such rulings do not conflict with national law.

President Paul Biya's Cameroon People's Democratic Movement (CPDM) was the only legal political party until December 1990. Numerous regional political groups have since formed. The primary opposition is the Social Democratic Front (SDF), based largely in the Anglophone region of the country and headed by John Fru Ndi. Biya and his party have maintained control of the presidency and the National Assembly in national elections, but rivals contend that these have been unfair. Human rights organisations allege that the government suppresses the freedoms of opposition groups by preventing demonstrations, disrupting meetings, and arresting opposition leaders and journalists. Freedom House ranks Cameroon as "not free" in terms of political rights and civil liberties. The last parliamentary elections were held on 22 July 2007.

Cameroon is a member of both the Commonwealth of Nations and La Francophonie. Its foreign policy closely follows that of its main ally, France (the former colonial ruler). The country relies heavily on France for its defence, although military spending is high in comparison to other sectors of government. Biya has clashed with the government of Nigeria over possession of the Bakassi peninsula and with Gabon's president, El Hadj Omar Bongo, over personal rivalries.

Education and Health

Most children have access to free, state-run schools or subsidised, private and religious facilities. The educational system is a mixture of British and French precedents with most instruction in English or French. Cameroon has one of the highest school attendance rates in Africa. Girls attend school less regularly than boys do because of cultural attitudes, domestic duties, early marriage and pregnancy, and sexual harassment. Although attendance rates are higher in the south, a disproportionate number of teachers are stationed there, leaving northern schools chronically understaffed.

Six state-run universities serve Cameroon's student population. More than 60,000 students were enrolled for the 1998–1999 school year. A council of deans, school directors, and representatives of state ministries governs the schools under the leadership of a vice-chancellor. State funding for universities is low, and student registrations nominally make up 25% of the higher education budget. However, students have fought these fees since their introduction in 1993. Universities have resisted the urge to increase the selectiveness of admissions in an effort to increase revenue from student fees, and the student populations have increased well beyond the 5,000 they were built to educate. Likewise, cuts in faculty salaries in 1993 made it difficult to find and keep qualified staff.

Since 1990, private institutions have sprung up in five regions. These schools charge fees that are five to ten times those levied by state schools. Nevertheless, they offer short professional-training programmes in areas such as accounting, management, journalism, and Internet technologies, so they are popular with students. Certain schools nonetheless fall short of government minimum standards of infrastructure and faculty and must operate unlicensed.

The quality of health care is generally low. Outside the major cities, facilities are often dirty and poorly equipped. Endemic diseases include dengue fever, filariasis, leishmaniasis, malaria, meningitis, schistosomiasis, and sleeping sickness. The HIV/AIDS seroprevalence rate is estimated at 5.4% for those aged 15–49, although a strong stigma against the illness keeps the number of reported cases artificially low. Traditional healers remain a popular alternative to Western medicine.

Territorial Administration

The Ministry of Territorial Administration (MINATD) is responsible for preparation, implementation and assessment of government policy on territorial administration, decentralization, and civil protection, which includes prevention of natural and planning for disasters. It is the duty of the MINATD to organize the territorial administrative units and external services and to organize national and local elections and referendums.

Regions

The constitution divides Cameroon into 10 semi-autonomous regions, each under the administration of an elected Regional Council. A presidential decree of 12 November 2008 officially instigated the change from provinces to regions. Each region is headed by a presidentially appointed governor. These leaders are charged with implementing the will of the president, reporting on the general mood and conditions of the regions, administering the civil service, keeping the peace, and overseeing the heads of the smaller administrative units. Governors have broad powers: they may order propaganda in their area and call in the army, gendarmes, and police. All local government officials are employees of the central government's Ministry of Territorial Administration, from which local governments also get most of their budgets.

The regions are subdivided into 58 divisions (French *departements*). These are headed by presidentially appointed divisional officers (*prefets*), who perform the governors' duties on a smaller scale. The divisions are further sub-divided into sub-divisions (*arrondissements*), headed by assistant divisional officers (*sous-prefets*). The districts, administered by district heads (*chefs de district*), are the smallest administrative units. These are found in large sub-divisions and in regions that are difficult to reach.

The three northernmost regions are the Far North (*Extreme Nord*), North (*Nord*), and Adamawa (*Adamaoua*). Directly south of them are the Centre (*Centre*) and East (*Est*). The South Province (*Sud*) lies on the Gulf of Guinea and the southern border. Cameroon's western region is split into four smaller regions: The Littoral (*Littoral*) and Southwest (*Sud-Ouest*) regions are on the coast, and the Northwest (*Nord-Ouest*) and West (*Ouest*) regions are in the western grassfields. The Northwest and Southwest were once part of British Cameroons; the other regions were in French Cameroun.

Geography and Climate

At 475,442 square kilometres (183,569 sq mi), Cameroon is the world's 53rd-largest country. It is comparable in size to Papua New Guinea and somewhat larger than the U.S. state of California. The country is located in Central and West Africa on the Bight of Bonny, part of the Gulf of Guinea and the Atlantic Ocean. Tourist literature describes Cameroon as "Africa in miniature" because it exhibits all major climates and vegetation of the continent: coast, desert, mountains, rainforest, and savanna. The country's neighbours are Nigeria to the west; Chad to the northeast; the Central African Republic to the east; and Equatorial Guinea, Gabon, and the Republic of the Congo to the south.

Cameroon is divided into five major geographic zones distinguished by dominant physical, climatic, and vegetative features. The coastal plain extends 15 to 150 kilometres (9 to 93 mi) inland from the Gulf of Guinea and has an average elevation of 90 metres (295 ft). Exceedingly hot and humid with a short dry season, this belt is densely forested and includes some of the wettest places on earth, part of the Cross-Sanaga-Bioko coastal forests.

The South Cameroon Plateau rises from the coastal plain to an average elevation of 650 metres (2,133 ft). Equatorial rainforest dominates this region, although its alternation between wet and dry seasons makes it is less humid than the coast. This area is part of the Atlantic Equatorial coastal forests ecoregion.

An irregular chain of mountains, hills, and plateaus known as the Cameroon range extends from Mount Cameroon on the coast—Cameroon's highest point at 4,095 metres (13,435 ft)—almost to Lake Chad at Cameroon's northern tip. This region has a mild climate, particularly on the Western High Plateau, although rainfall is high. Its soils are among Cameroon's most fertile, especially around volcanic Mount Cameroon. Volcanism here has created crater lakes. On 21 August 1986, one of these, Lake Nyos, belched carbon dioxide and killed between 1,700 and 2,000 people. This area has been delineated by the World Wildlife Fund as the Cameroonian Highlands forests ecoregion.

The southern plateau rises northward to the grassy, rugged Adamawa Plateau. This feature stretches from the western mountain area and forms a barrier between the country's north and south. Its average elevation is 1,100 metres (3,609 ft), and its temperature ranges from 22 °C (71.6 °F) to 25 °C (77 °F) with high rainfall. The northern lowland region extends from the edge of the Adamawa to Lake Chad with an average elevation of 300 to 350 metres (984 to 1,148 ft). Its characteristic vegetation is savanna scrub and grass. This is an arid region with sparse rainfall and high median temperatures.

Cameroon has four patterns of drainage. In the south, the principal rivers are the Ntem, Nyong, Sanaga, and Wouri. These flow southwestward or westward directly into the Gulf of Guinea. The Dja and Kadei drain southeastward into the

Congo River. In northern Cameroon, the Benoue River runs north and west and empties into the Niger. The Logone flows northward into Lake Chad, which Cameroon shares with three neighbouring countries.

Economy and Infrastructure

Cameroon's per-capita GDP (PPP) was estimated as US $2,300 in 2008, one of the ten highest in sub-Saharan Africa. Major export markets include France, Italy, South Korea, Spain, and the United Kingdom. Cameroon is part of the Bank of Central African States (of which it is the dominant economy), the Customs and Economic Union of Central Africa (UDEAC) and the Organization for the Harmonization of Business Law in Africa (OHADA).

Its currency is the CFA franc. Red tape, high taxes, and endemic corruption have impeded growth of the private sector. Unemployment was estimated at 30% in 2001, and about a third of the population was living below the international poverty threshold of US$1.25 a day in 2009. Since the late 1980s, Cameroon has been following programmes advocated by the World Bank and International Monetary Fund (IMF) to reduce poverty, privatise industries, and increase economic growth. Tourism is a growing sector, particularly in the coastal area, around Mount Cameroon, and in the north.

Cameroon's natural resources are very well suited to agriculture and arboriculture. An estimated 70% of the population farms, and agriculture comprised an estimated 19.8% of GDP in 2009. Most agriculture is done at the subsistence scale by local farmers using simple tools. They sell their surplus produce, and some maintain separate fields for commercial use. Urban centres are particularly reliant on peasant agriculture for their foodstuffs. Soils and climate on the coast encourage extensive commercial cultivation of bananas, cocoa, oil palms, rubber, and tea. Inland on the South Cameroon Plateau, cash crops include coffee, sugar, and tobacco. Coffee is a major cash crop in the western highlands, and in the north, natural conditions favour crops such as cotton, groundnuts, and rice. Reliance on agricultural exports makes Cameroon vulnerable to shifts in their prices.

Livestock are raised throughout the country. Fishing employs some 5,000 people and provides 20,000 tons of seafood each year. Bushmeat, long a staple food for rural Cameroonians, is today a delicacy in the country's urban centres. The commercial bushmeat trade has now surpassed deforestation as the main threat to wildlife in Cameroon.

The southern rainforest has vast timber reserves, estimated to cover 37% of Cameroon's total land area. However, large areas of the forest are difficult to reach. Logging, largely handled by foreign-owned firms, provides the government US$60 million a year, and laws mandate the safe and sustainable exploitation of timber. Nevertheless, in practice, the industry is one of the least regulated in Cameroon.

Factory-based industry accounted for an estimated 29.7% of GDP in 2009. More than 75% of Cameroon's industrial strength is located in Douala and Bonaberi. Cameroon possesses substantial mineral resources, but these are not extensively mined. Petroleum exploitation has fallen since 1985, but this is still a substantial sector such that dips in prices have a strong effect on the economy.

Rapids and waterfalls obstruct the southern rivers, but these sites offer opportunities for hydroelectric development and supply most of Cameroon's energy. The Sanaga River powers the largest hydroelectric station, located at Edea. The rest of Cameroon's energy comes from oil-powered thermal engines. Much of the country remains without reliable power supplies.

Transport in Cameroon is often difficult. Except for the several relatively good toll roads which connect major cities (all of them one-lane) roads are poorly maintained and subject to inclement weather, since only 10% of the roadways are tarred. Roadblocks often serve little other purpose than to allow police and gendarmes to collect bribes from travellers. Road banditry has long hampered transport along the eastern and western borders, and since 2005, the problem has intensified in the east as the Central African Republic has further destabilised.

Intercity bus services run by multiple private companies connect all major cities. Although intercity buses rarely depart on schedule but rather wait until all the tickets are sold. They are the most popular mean of transportation followed by the government-owned rail service. Rail service runs from Kumba in the west to Belabo in the east and north to Ngaoundere. International airports are located in Douala and Garoua with a smaller facility at Yaounde.

The airport at Bamenda is now closed. The Wouri River estuary provides a harbour for Douala, the country's principal seaport. In the north, the Benoue River is seasonally navigable from Garoua across into Nigeria.

Although press freedoms have improved since the early 2000s, the press is corrupt and beholden to special interests and political groups. Newspapers routinely self-censor to avoid government reprisals. The major radio and television stations are state-run and other communications, such as land-based telephones and telegraphs, are largely under government control. However, cell phone networks and Internet providers have increased dramatically since the early 2000s and are largely unregulated.

Demographics

2009 UN estimates place Cameroon's population at 19,522,000. The population is young: an estimated 40.9% are under 15, and 96.7% are under 65. The birth rate is estimated at 34.1 births per 1,000 people, the death rate at 12.2. The life expectancy is 53.69 years (52.89 years for males and 54.52 years for females).

Cameroon's population is almost evenly divided between urban and rural dwellers. Population density is highest in the large urban centres, the western highlands, and the northeastern plain. Douala, Yaounde, and Garoua are the largest cities. In contrast, the Adamawa Plateau, southeastern Benoue depression, and most of the South Cameroon Plateau are sparsely populated.

People from the overpopulated western highlands and the underdeveloped north are moving to the coastal plantation zone and urban centres for employment. Smaller movements are occurring as workers seek employment in lumber mills and plantations in the south and east. Although the national sex ratio is relatively even, these out-migrants are primarily males, which leads to unbalanced ratios in some regions. Both monogamous and polygamous marriage are practiced, and the average Cameroonian family is large and extended. In the north, women tend to the home, and men herd cattle or work as farmers. In the south, women grow the family's food, and men provide meat and grow cash crops. Cameroonian society is male-dominated, and violence and discrimination against women is common.

Estimates identify anywhere from 230 to 282 different folks and linguistic groups in Cameroon. The Adamawa Plateau broadly bisects these into northern and southern divisions. The northern peoples are Sudanese groups, who live in the central highlands and the northern lowlands, and the Fulani, who are spread throughout northern Cameroon.

A small number of Shuwa Arabs live near Lake Chad. Southern Cameroon is inhabited by speakers of Bantu and Semi-Bantu languages. Bantu-speaking groups inhabit the coastal and equatorial zones, while speakers of Semi-Bantu languages live in the Western grassfields. Some 5,000 Pygmies roam the southeastern and coastal rainforests or live in small, roadside settlements. Nigerians, make up the largest group of foreign nationals.

In 2007, Cameroon hosted a total population of refugees and asylum seekers of approximately 97,400. Of these, 49,300 were from the Central African Republic (many driven west by war), 41,600 from Chad, and 2,900 from Nigeria. Kidnappings of Cameroonian citizens by Central African bandits have increased since 2005.

The European languages introduced during colonialism have created a linguistic divide between the population who live in the Northwest and Southwest regions and the French-speaking remainder of the country.

Both English and French are official languages, although French is by far the most understood language. Cameroonian Pidgin English is the lingua franca in the formerly British-administered territories. A mixture of English, French, and Pidgin called Camfranglais has been gaining popularity in urban centres since the mid-1970s.

Religion

Christian	40%
Indigenous	40%
Islam	20%

Cameroon has a high level of religious freedom and diversity. Christians are concentrated chiefly in the southern and western regions, and Muslims reside in large numbers in every region but are concentrated in the north. There is significant internal migration. There are currently no active Islamic political parties. Large cities have significant populations of both groups, with mosques and churches often located near each other.

People from the North-West and South-West provinces are largely Protestant, and the French speaking regions of the southern and western regions are largely Catholic. Southern ethnic groups predominantly follow Christian or Animist beliefs, or a syncretic combination of the two. People widely believe in witchcraft, and the government outlaws such practices. Suspected witches are often subject to mob violence.

In the northern regions, the locally dominant Fulani (or Peuhl) ethnic group is mostly Muslim, although some ethnic groups retain native animist beliefs and are called *Kirdi* ("pagan") by the Fulani. The Bamum ethnic group of the West Region is largely Muslim. Native Traditional religions are practiced in rural areas throughout the country but rarely are practiced publicly in cities, in part because many indigenous religious groups are intrinsically local in character.

Culture

Each of Cameroon's ethnic groups has its own unique cultural forms. Typical celebrations include births, deaths, plantings, harvests, and religious rituals. Seven national holidays are observed throughout the year, and movable holidays include the Christian holy days of Good Friday, Easter Sunday, Easter Monday, and Ascension; and the Muslim holy days of Eid-ul-Fitr, Eid-ul-Adha, and Eid Miladun Nabi.

Music and dance are an integral part of Cameroonian ceremonies, festivals, social gatherings, and storytelling. Traditional dances are highly choreographed and separate men and women or forbid participation by one sex altogether. The goals of dances range from pure entertainment to religious devotion. Traditionally, music is transmitted orally. In a typical performance, a chorus of singers echoes a soloist.

Musical accompaniment may be as simple as clapping hands and stomping feet, but traditional instruments include bells worn by dancers, clappers, drums and talking drums, flutes, horns, rattles, scrapers, stringed instruments, whistles, and xylophones; the exact combination varies with ethnic group and region. Some performers sing complete songs by themselves, accompanied by a harplike instrument.

Popular music styles include ambasse bey of the coast, assiko of the Bassa, mangambeu of the Bangangte, and tsamassi of the Bamileke. Nigerian music has influenced Anglophone Cameroonian performers, and Prince Nico Mbarga's highlife hit "Sweet Mother" is the top-selling African record in history. The two most popular styles are makossa and bikutsi. Makossa developed in Douala and mixes folk music, highlife, soul, and Congo music. Performers such as Manu Dibango, Francis Bebey, Moni Bile, and Petit-Pays popularised the style worldwide in the 1970s and 1980s. Bikutsi originated as war music among the Ewondo. Artists such as Anne-Marie Nzie developed it into a popular dance music beginning in the 1940s, and performers such as Mama Ohandja and Les Tetes Brulees popularised it internationally during the 1960s, 1970s, and 1980s.

Cuisine varies by region, but a large, one-course, evening meal is common throughout the country. A typical dish is based on cocoyams, maize, cassava (manioc), millet, plantains, potatoes, rice, or yams, often pounded into dough-like fufu (*cous-cous*). This is served with a sauce, soup, or stew made from greens, groundnuts, palm oil, or other ingredients. Meat and fish are popular but expensive additions. Dishes are often quite hot, spiced with salt, red pepper, and Maggi. Water, palm wine, and millet beer are the traditional mealtime drinks, although beer, soda, and wine have gained popularity. Silverware is common, but food is traditionally manipulated with the right hand. Breakfast consists of leftovers of bread and fruit with coffee or tea, generally breakfast is made from wheat flour various different foods such as puff-puff (doughnuts), accra banana made from bananas and flour, bean cakes and many more. Snacks are popular, especially in larger towns where they may be bought from street vendors.

Traditional arts and crafts are practiced throughout the country for commercial, decorative, and religious purposes. Woodcarvings and sculptures are especially common. The high-quality clay of the western highlands is suitable for pottery and ceramics. Other crafts include basket weaving, beadworking, brass and bronze working, calabash carving and painting, embroidery, and leather working. Traditional housing styles make use of locally available materials and vary from temporary wood-and-leaf shelters of nomadic Mbororo to the rectangular mud-and-thatch homes of southern peoples. Dwellings made from materials such as cement and tin are increasingly common.

Cameroonian literature and film have concentrated on both European and African themes. Colonial-era writers such as Louis-Marie Pouka and Sankie Maimo were educated by European missionary societies and advocated assimilation into European culture as the means to bring Cameroon into the modern world. After World War II, writers such as Mongo Beti and Ferdinand Oyono analysed and criticised colonialism and rejected assimilation.

Shortly after independence, filmmakers such as Jean-Paul Ngassa and Therese Sita-Bella explored similar themes. In the 1960s, Mongo Beti and other writers explored post-colonialism, problems of African development, and the recovery of African identity. Meanwhile, in the mid-1970s, filmmakers such as Jean-Pierre Dikongue Pipa and Daniel Kamwa dealt with the conflicts between traditional and post-colonial society. Literature and films during the next two decades concentrated more on wholly Cameroonian themes.

National policy strongly advocates sport in all forms. Traditional sports include canoe racing and wrestling, and several hundred runners participate in the 40 km (25 mi) Mount Cameroon Race of Hope each year. Cameroon is one of the few tropical countries to have competed in the Winter Olympics. However, sport in Cameroon is dominated by association football (soccer). Amateur football clubs abound, organised along ethnic lines or under corporate sponsors. The Cameroon national football team has been one of the most successful in the world since its strong showing in the 1990 FIFA World Cup. Cameroon has won four African Cup of Nations titles and the gold medal at the 2000 Olympics.

Industrial Zones/regions

- Douala industrial region
  - Bonaberi Industrial Zone
  - Dibombari industrial zone
  - Bassa industrial zone.
- Edea industrial region
- Tiko-Limbe Industrial region
- Yaounde industrial region
- Garoua – Figuil region
- Befoussam – Bamenda industrial region.

12

The Republic of Mali

Mali, officially the Republic of Mali (French: *Republique du Mali*), is a landlocked country in Western Africa. Mali borders Algeria on the north, Niger on the east, Burkina Faso and the Cote d'Ivoire on the south, Guinea on the south-west, and Senegal and Mauritania on the west. Its size is just over 1,240,000 km^2 with a population more than 14 million. Its capital is Bamako.

Mali consists of eight regions and its borders on the north reach deep into the middle of the Sahara, while the country's southern region, where the majority of inhabitants live, features the Niger and Senegal rivers.

The country's economic structure centres around agriculture and fishing. Some of Mali's natural resources include gold, uranium, and salt. Mali is considered to be one of the poorest nations in the world.

Present-day Mali was once part of three West African empires that controlled trans-Saharan trade: the Ghana Empire, the Mali Empire (from which Mali is named), and the Songhai Empire. In the late 1800s, Mali fell under French control, becoming part of French Sudan. Mali gained independence in 1959 with Senegal, as the Mali Federation. A year later, the Mali Federation became the independent nation of Mali. After a long period of one-party rule, a 1991 coup led to the writing of a new constitution and the establishment of Mali as a democratic, multi-party state. About half the population live below the international poverty line of US$1.25 a day.

History

Mali was once part of three famed West African empires which controlled trans-Saharan trade in gold, salt, slaves, and other precious commodities. These Sahelian kingdoms had neither rigid geopolitical boundaries nor rigid ethnic identities. The earliest of these empires was the Ghana Empire, which was dominated by the Soninke, a Mande-speaking people. The nation expanded throughout West Africa from the 8th century until 1078, when it was conquered by the Almoravids.

The Mali Empire later formed on the upper Niger River, and reached the height of power in the fourteenth century. Under the Mali Empire, the ancient cities of Djenne and Timbuktu were centres of both trade and Islamic learning. The empire later declined as a result of internal intrigue, ultimately being supplanted by the Songhai Empire. The Songhai people originated in current northwestern Nigeria. The Songhai had long been a major power in West Africa subject to the Mali Empire's rule.

In the late 14th century, the Songhai gradually gained independence from the Mali Empire and expanded, ultimately subsuming the entire eastern portion of the Mali Empire. The Songhai Empire's eventual collapse was largely the result of a Moroccan invasion in 1591, under the command of Judar Pasha. The fall of the Songhai Empire marked the end of the region's role as a trading crossroads. Following the establishment of sea routes by the European powers, the trans-Saharan trade routes lost significance. The worst recorded famine occurred between 1738 and 1756, killing about half of the population of Timbuktu.

In the colonial era, Mali fell under the control of the French beginning in the late 19th century. By 1905, most of the area was under firm French control as a part of French Sudan. In early 1959, Mali (then the Sudanese Republic) and Senegal united to become the Mali Federation. The Mali Federation gained independence from France on June 20, 1960. Senegal withdrew from the federation in August 1960, which allowed the Sudanese Republic to form the independent nation of Mali on September 22, 1960. Modibo Keita was elected the first president. Keita quickly established a one-party state, adopted an independent African and socialist orientation with close ties to the East, and implemented extensive nationalization of economic resources.

In November 1968, following progressive economic decline, the Keita regime was overthrown in a bloodless military coup led by Moussa Traore. The subsequent military-led regime, with Traore as president, attempted to reform the economy. However, his efforts were frustrated by political turmoil and a devastating drought between 1968 to 1974, which killed thousands of people from famine. The Traore regime faced student unrest beginning in the late 1970s and three coup attempts. However, the Traore regime repressed all dissenters until the late 1980s.

The government continued to attempt economic reforms, and the populace became increasingly dissatisfied. In response to growing demands for multi-party democracy, the Traore regime allowed some limited political liberalization, but refused to usher in a full-fledged democratic system. In 1990, cohesive opposition movements began to emerge, and was complicated by the turbulent rise of ethnic violence in the north following the return of many Tuaregs to Mali.

Anti-government protests in 1991 led to a coup, a transitional government, and a new constitution. In 1992, Alpha Oumar Konare won Mali's first democratic, multi-party presidential election. Upon his re-election in 1997, President Konare pushed through political and economic reforms and fought corruption. In 2002, he was succeeded in democratic elections by Amadou Toumani Toure, a retired general, who had been the leader of the military aspect of the 1991 democratic uprising. Today, Mali is one of the most politically and socially stable countries in Africa.

Geography

Mali is a landlocked nation in West Africa, located southwest of Algeria. At 1,240,000 square kilometres (478,767 sq mi), Mali is the world's 24th-largest country and is comparable in size to South Africa or Angola. Most of the country lies in the southern Sahara, which produces a hot, dust-laden Sudanian savanna zone. Mali is mostly flat, rising to rolling northern plains covered by sand. The Adrar des Ifoghas lies in the northeast.

The country's climate ranges from tropical in the south to arid in the north. Most of the country receives negligible rainfall; droughts are frequent. Late June to early December is the rainy season. During this time, flooding of the Niger River is common, creating the Inner Niger Delta. The nation has considerable natural resources, with gold, uranium, phosphates, kaolinite, salt and limestone being most widely exploited. Mali faces numerous environmental challenges, including desertification, deforestation, soil erosion, and inadequate supplies of potable water.

Regions and Cercles

Mali is divided into eight regions (*regions*) and one district. Each region has a governor. Since Mali's regions are very large, the country is subdivided into 49 cercles, totalling 288 arrondissements. Mayors and elected members of the city councils officiate the arrondissements.

The regions and districts are:

- Gao
- Kayes
- Kidal
- Koulikoro
- Mopti
- Segou
- Sikasso
- Tombouctou (Timbuktu)
- Bamako (capital district)

Politics and Government

Mali is a constitutional democracy governed by the constitution of January 12, 1992, which was amended in 1999. The constitution provides for a separation of powers among the executive, legislative, and judicial branches of government. The system of government can be described as "semi-presidential."

Executive power is vested in a president, who is elected to a five-year term by universal suffrage and is limited to two terms. The president serves as a chief of state and commander in chief of the armed forces. A prime minister appointed by the president serves as head of government and in turn appoints the Council of Ministers. The unicameral National Assembly is Mali's sole legislative body, consisting of deputies elected to five-year terms. Following the 2007 elections, the Alliance for Democracy and Progress held 113 of 160 seats in the assembly. The assembly holds two regular sessions each year, during which it debates and votes on legislation that has been submitted by a member or by the government. Democracy-wise things looked positive after the local elections at the end of April 2009, though significant shortcomings and attempts at manipulation still existed.

Mali's constitution provides for an independent judiciary, but the executive continues to exercise influence over the judiciary by virtue of power to appoint judges and oversee both judicial functions and law enforcement. Mali's highest courts are the Supreme Court, which has both judicial and administrative powers, and a separate Constitutional Court that provides judicial review of legislative acts and serves as an election arbiter. Various lower courts exist, though village chiefs and elders resolve most local disputes in rural areas.

Foreign Relations and Military

Mali's foreign policy orientation has become increasingly pragmatic and pro-Western over time. Since the institution of a democratic form of government in 2002, Mali's relations with the West in general and with the United States in particular have improved significantly. Mali has a longstanding yet ambivalent relationship with France, a former colonial ruler. Mali is active in regional organizations such as the African Union. Working to control and resolve regional conflicts, such as in Cote d'Ivoire, Liberia, and Sierra Leone, is one of Mali's major foreign policy goals. Mali feels threatened by the potential for the spillover of conflicts in neighbouring states, and relations with those neighbours are often uneasy. General insecurity along borders in the north, including cross-border banditry and terrorism, remain troubling issues in regional relations. Mali's military forces consist of an army, which includes land forces and air force, as well as the paramilitary Gendarmerie and Republican Guard, all of which are under the control of Mali's Ministry of Defence and Veterans, headed by a civilian. The military is underpaid, poorly equipped, and in need of rationalization. Organization has suffered from the incorporation of Tuareg irregular forces into the regular military following a 1992 agreement between the government and Tuareg rebel forces. The military has generally kept a low profile since the democratic transition of 1992. The incumbent president, Amadou Toumani Toure, is a former army general and as such reportedly enjoys widespread military support. In the annual human rights report for 2003, the U.S. Department of State rated civilian control of security forces as generally effective but noted a few "instances in which elements of the security forces acted independently of government authority.". Western powers such as the United States have also helped Mali's military with training and equipment.

Economy

Mali is one of the poorest countries in the world. The average worker's annual salary is approximately US$1,500. Between 1992 and 1995, Mali implemented an economic adjustment program that resulted in economic growth and a reduction in financial imbalances. The program increased social and economic conditions, and led to Mali joining the World Trade Organization on May 31, 1995. The gross domestic product (GDP) has risen since. In 2002, the GDP amounted to US$3.4 billion, and increased to US$5.8 billion in 2005, which amounts to an approximately 17.6% annual growth rate.

Mali's key industry is agriculture. Cotton is the country's largest crop export and is exported west throughout Senegal and the Ivory Coast. During 2002, 620,000 tons of cotton were produced in Mali but cotton prices declined significantly in 2003. In addition to cotton, Mali produces rice, millet, corn, vegetables, tobacco, and tree crops. Gold, livestock and agriculture amount to eighty percent of Mali's exports. Eighty percent of Malian workers are employed in agriculture while fifteen percent work in the service sector. However, seasonal variations lead to regular temporary unemployment of agricultural workers. Mali's resource in livestock consists of millions of cattle, sheep, and goats. Approximately 40% of Mali's herds were lost during the Sahel drought in 1972-74.

In 1991, with the assistance of the International Development Association, Mali relaxed the enforcement of mining codes which led to renewed foreign interest and investment in the mining industry. Gold is mined in the southern region and Mali has the third highest gold production in Africa (after South Africa and Ghana). The emergence of gold as Mali's leading export product since 1999 has helped mitigate some of the negative impact of the cotton and Cote d'Ivoire crises. Other natural resources include kaolin, salt, phosphate, and limestone.

Electricity and water are maintained by the Energie du Mali, or EDM, and textiles are generated by Industry Textile du Mali, or ITEMA. Mali has made efficient use of hydroelectricity, consisting of over half of Mali's electrical power. In 2002, 700 GWh of hydroelectric power were produced in Mali.

The Malian government participates in foreign involvement, concerning commerce and privatization. Mali underwent economic reform, beginning in 1988 by signing agreements with the World Bank and the International Monetary Fund. During 1988 to 1996, Mali's government largely reformed public enterprises. Since the agreement, sixteen enterprises were privatized, twelve partially privatized, and twenty liquidated. In 2005, the Malian government conceded a railroad company to the Savage Corporation. Two major companies, Societe de Telecommunications du Mali (SOTELMA) and the Cotton Ginning Company (CMDT), are expected to be privatized in 2008.

Mali is a member of the Organization for the Harmonization of Business Law in Africa (OHADA).

Demographics

In July 2009, Mali's population was an estimated 13 million, with an annual growth rate of 2.7%. The population is predominantly rural (68% in 2002), and 5–10% of Malians are nomadic. More than 90% of the population lives in the southern part of the country, especially in Bamako, which has over 1 million residents. In 2007, about 48% of Malians were less than fifteen years old, 49% were 15–64 years old, and 3% were 65 and older. The median age was 15.9 years. The birth rate in 2007 was 49.6 births per 1,000, and the total fertility rate was 7.4 children per woman. The death rate in 2007 was 16.5 deaths per 1,000. Life expectancy at birth was 49.5 years total (47.6 for males and 51.5 for females). Mali has one of the world's highest rates of infant mortality, with 106 deaths per 1,000 live births in 2007.

Mali's population encompasses a number of sub-Saharan ethnic groups, most of which have historical, cultural, linguistic, and religious commonalities. The Bambara are by far the largest single ethnic group, making up 36.5% of the population. Collectively, the Bambara, Soninke, Khassonke, and Malinke, all part of the broader Mande group, constitute 50% of Mali's population. Other significant groups are the Peul (17%), Voltaic (12%), Songhai (6%), and Tuareg and Moor (10%). Mali historically has enjoyed reasonably good inter-ethnic relations; however, some hereditary servitude relationships exist, as do ethnic tensions between the Songhai and the Tuareg. Over the past 40 years, persistent drought has forced many Tuareg to give up their nomadic way of life.

Mali's official language is French, but numerous (40 or more) African languages also are widely used by the various ethnic groups. About 80% of Mali's population can communicate in Bambara, which is the country's principal lingua franca and marketplace language.

Religion

Religion in Mali

Religion	***Percent***
Islam	90%
Christianity	5%
Indigenous	5%

An estimated 90% of Malians are Muslim (mostly Sunni and Sufi), approximately 5% are Christian (about two-thirds Roman Catholic and one-third Protestant) and the remaining

5% adhere to indigenous or traditional animist beliefs. Atheism and agnosticism are believed to be rare among Malians, most of whom practice their religion on a daily basis. Islam as practiced in Mali is moderate, tolerant, and adapted to local conditions; relations between Muslims and practitioners of minority religious faiths are generally amicable. The constitution establishes a secular state and provides for freedom of religion, and the government largely respects this right.

Health and Education

Mali faces numerous health challenges related to poverty, malnutrition, and inadequate hygiene and sanitation. Mali's health and development indicators rank among the worst in the world. In 2000, only 62–65 percent of the population was estimated to have access to safe drinking water and only 69 percent to sanitation services of some kind.

In 2001, the general government expenditures on health totalled about US$4 per capita at an average exchange rate. Medical facilities in Mali are very limited, and medicines are in short supply. Malaria and other arthropod-borne diseases are prevalent in Mali, as are a number of infectious diseases such as cholera and tuberculosis. Mali's population also suffers from a high rate of child malnutrition and a low rate of immunization. An estimated 1.9 percent of the adult and children population was afflicted with HIV/AIDS that year, among the lowest rates in Sub-Saharan Africa.

Public education in Mali is in principle provided free of charge and is compulsory for nine years between the ages of seven and 16. The system encompasses six years of primary education beginning at age seven, followed by six years of secondary education. However, Mali's actual primary school enrollment rate is low, in large part because families are unable to cover the cost of uniforms, books, supplies, and other fees required to attend. In the 2000–01 school year, the primary school enrollment rate was 61% (71% of males and 51% of females); in the late 1990s, the secondary school enrollment rate was 15% percent (20% of males and 10% of females). The education system is plagued by a lack of schools in rural areas, as well as shortages of teachers and materials. Estimates of literacy rates in Mali range from 27–30% to 46.4%, with literacy rates significantly lower among women than men.

Culture

Malian musical traditions are derived from the griots, who are known as "Keeper of Memories". Malian music is diverse and has several different genres. Some famous Malian influences in music are kora virtouso musician Toumani Diabate, the late roots and blues guitarist Ali Farka Toure, the Tuareg band Tinariwen, and several Afro-pop artists such as Salif Keita, the duo Amadou et Mariam, Oumou Sangare, and Habib Koite.

Though Mali's literature is less famous than its music, Mali has always been one of Africa's liveliest intellectual centres. Mali's literary tradition is passed mainly by word of mouth, with *jalis* reciting or singing histories and stories known by heart. Amadou Hampate Ba, Mali's best-known historian, spent much of his life writing these oral traditions down for the world to remember. The best-known novel by a Malian writer is Yambo Ouologuem's *Le devoir de violence*, which won the 1968 Prix Renaudot but whose legacy was marred by accusations of plagiarism. Other well-known Malian writers include Baba Traore, Modibo Sounkalo Keita, Massa Makan Diabate, Moussa Konate, and Fily Dabo Sissoko.

The varied everyday culture of Malians reflects the country's ethnic and geographic diversity. Most Malians wear flowing, colourful robes called boubous that are typical of West Africa. Malians frequently participate in traditional festivals, dances, and ceremonies. Rice and millet are the staples of Malian cuisine, which is heavily based on cereal grains. Grains are generally prepared with sauces made from leaves such spinach or baobab leaves, with tomato, or with peanut sauce, and may be accompanied by pieces of grilled meat (typically chicken, mutton, beef, or goat). Malian cuisine varies regionally.

Sport

The most popular sport in Mali is football (soccer), which became more prominent after Mali hosted the 2002 African Cup of Nations. Most towns have regular games; the most popular teams nationally are Djoliba AC, Stade Malien, and

Real Bamako, all based in the capital. Informal games are often played by youths using a bundle of rags as a ball. The country has produced several notable players for French teams, including Salif Keita and Jean Tigana. Frederic "Fredi" Kanoute, named 2007 African Footballer of the Year, currently plays for Sevilla FC in Spain's La Liga. Also playing for major clubs in Spain are Mahamadou Diarra, captain of the Mali national squad, for Real Madrid and Seydou Keita for FC Barcelona. Other notable players currently on European squads include, Mamady Sidibe (Stoke City), Mohammed Sissoko (Juventus), Sammy Traore (Paris Saint-Germain), Adama Coulibaly (AJ Auxerre), Kalifa Cisse and Jimmy Kebe (Reading F.C.), and Dramane Traore (Lokomotiv Moscow). Basketball is another major sport; the Mali women's national basketball team, led by Hamchetou Maiga, competed at the 2008 Beijing Olympics. Traditional wrestling (*la lutte*) is also somewhat common, though popularity has declined in recent years. The game wari, a mancala variant, is a common pastime.

13

Sierra Leone

Sierra Leone officially the Republic of Sierra Leone, is a country in West Africa. It is bordered by Guinea in the north, Liberia in the southeast, and the Atlantic Ocean in the southwest. Sierra Leone covers a total area of 71,740 km (27,699 sq mi) and has a population estimated at 5.2 million. The country is a former British Colony and is now a constitutional republic comprising three provinces and the Western Area, which are further divided into fourteen districts.

The country has a tropical climate, with a diverse environment ranging from savannah to rainforests. Freetown is the capital, largest city and economic centre. The other major cities are Bo, Kenema, Koidu Town and Makeni. English is the official language, spoken at schools, government administration and by the media. However, the Krio language (a language derived from English and several African languages and native to the Sierra Leone Krio people) is the most widely spoken language in virtually all parts of the country. The Krio language is spoken by 97% of the country's population and unites all the different ethnic groups, especially in their trade and interaction with each other. Despite its common use throughout the country, the Krio language has no official status.

Sierra Leone is very rich in mineral resources, possessing most of the known mineral types of the world, many of which are found in significant quantities. The country has relied on mining, especially diamonds, for its economic base. The country is among the top 10 diamond producing nations in the world, and mineral exports remain the main foreign currency earner.

Sierra Leone is among the largest titanium and bauxite producing nations on the planet. Sierra Leone is also a major producer of gold.

Sierra Leone has one of the world's largest deposits of rutile, a titanium ore used in paint pigment and welding rod coatings. In September 2009, Anadarko Petroleum Corporation, a United States oil firm, announced the discovery of an "active petroleum deepwater oil well" off the coast of Sierra Leone. Despite being rich in mineral resources, the majority of its people live in poverty.

Sierra Leone is also home to the third largest natural harbour in the world, the Queen Elizabeth II Quay (also known as the QE II Quay and locally as the Deep Water Quay or Government Wharf).

Early inhabitants of Sierra Leone included the Sherbro, Temne and Limba, and Tyra peoples, and later the Mende, who knew the country as Romarong, and the Kono who settled in the East of the country. In 1462, it was visited by the Portuguese explorer Pedro da Cintra, who gave dubbed it Serra de Leao, meaning "Lion Mountains". Sierra Leone later became an important centre of the transatlantic trade in slaves until 1792 when Freetown was founded by the Sierra Leone Company as a home for formerly enslaved African Americans. In 1808, Freetown became a British Crown Colony, and in 1896, the interior of the country became a British Protectorate; in 1961, the two combined and gained independence.

Over two decades of government neglect of the interior followed by the spilling over of the Liberian conflict into its borders eventually led to the Sierra Leone Civil War, which began in 1991 and was resolved in 2000 after the struggling Nigerian-led United Nations troops were heavily reinforced by a British force spearheaded by 42 Commando of the Royal Marines as well as several British Army units. The arrival of this force in what was codenamed Operation Palliser resulted in the defeat of rebel forces and restored the civilian government elected in 1998 to Freetown. Since then, almost 72,500 former combatants have been disarmed and the country has re-established a functioning democracy. The Special Court for Sierra Leone was set up in 2002 to deal with war crimes and crimes against humanity committed since 1996. Sierra Leone is the third-lowest-ranked country on the Human Development Index and eighth-lowest on the Human Poverty Index, suffering from endemic corruption and suppression of the press.

History

Early History

Archaeological finds show that Sierra Leone has been inhabited continuously for at least 2,500 years, populated by successive movements from other parts of Africa. The use of iron was introduced to Sierra Leone by the 9th century, and by AD 1000 agriculture was being practiced by coastal tribes. Sierra Leone's dense tropical rainforest largely protected it from the influence of any precolonial African empires and from further Islamic colonization, which were unable to penetrate through it until the 18th century.

European contacts with Sierra Leone were among the first in West Africa. In 1462, Portuguese explorer Pedro da Cintra mapped the hills surrounding what is now Freetown Harbour, naming shaped formation *Serra de Leao* (Portuguese for *Lion Mountains*). The Italian rendering of this geographic formation is *Sierra Leone*, which became the country's name. Soon after Portuguese traders arrived at the harbour and by 1495 a fort that acted as a trading post had been built. The Portuguese were joined by the Dutch and French; all of them using Sierra Leone as a trading point for slaves. In 1562 the English joined the trade in human beings when Sir John Hawkins shipped 300 enslaved people, he had acquired 'by the sword and partly by other means', to the new colonies in America.

Enslavement and Freedom

In 1787, a plan was established to settle some of London's "Black Poor" in Sierra Leone in what was called the "Province of Freedom". A number of "Black Poor" arrived off the coast of Sierra Leone on May 15, 1787, accompanied by some English tradesmen. This was organized by the St. George's Bay Company, composed of British philanthropists who preferred it as a solution to continuing to financially support them in London. Many of the "Black poor" were African Americans, who had been promised their freedom for joining the British Army during the American Revolution, but also included other African and Asian inhabitants of London.

Disease and hostility from the indigenous people nearly eliminated the first group of colonists. Through intervention by Thomas Peters, the Sierra Leone Company was established to relocate another group of formerly enslaved Africans, this time nearly 1,200 Black Nova Scotians, most of whom had escaped enslavement in the United States. Given the most barren land in Nova Scotia, many had died from the harsh winters there. They established a settlement at Freetown in 1792 led by Peters. It was joined by other groups of freed Africans and became the first African-American haven for formerly enslaved Africans.

Though the English abolitionist Granville Sharp originally planned Sierra Leone as a utopian community, the directors of the Sierra Leone Company refused to allow the settlers to take freehold of the land. Knowing how Highland Clearances benefited Scottish landlords but not tenants, the settlers revolted in 1799. The revolt was only put down by the arrival of over 500 Jamaican Maroons, who also arrived via Nova Scotia.

Thousands of formerly enslaved Africans were returned to or liberated in Freetown. Most chose to remain in Sierra Leone. These returned Africans were from many areas of Africa, but principally the west coast. They joined the previous settlers and together became known as Creole or Krio people.

Cut off from their homes and traditions, they assimilated some aspects of British styles of inhabitants and built a flourishing trade of flowers and beads on the West African coast. The lingua franca of the colony was Krio, a creole language rooted in 18th century African American English, which quickly spread across the region as a common language of trade and Christian mission. In the 1790s, blacks voted for the first time in elections, as did women.

Colonial Era

In the early 20th century, Freetown served as the residence of the British governor who also ruled the Gold Coast (now Ghana) and the Gambia settlements. Sierra Leone also served as the educational centre of British West Africa. Fourah Bay College, established in 1827, rapidly became a magnet for English-speaking Africans on the West Coast. For more than a century, it was the only European-style university in western Sub-Saharan Africa.

During Sierra Leone's colonial history, indigenous people mounted several unsuccessful revolts against British rule and Krio domination. The most notable was the Hut Tax war of 1898. Its first leader was Bai Bureh, a Temne chief who refused to recognize the British-imposed tax on "huts" (dwellings). The tax was generally regarded by the native chiefs as an attack on their sovereignty. After the British issued a warrant to arrest Bai Bureh alleging that he had refused to pay taxes, he brought fighters from several Temne villages under his command, and from Limba, Loko, Soso, Kissi, and Mandinka villages.

Bureh's fighters had the advantage over the vastly more powerful British for several months of the war. Hundreds of British troops and hundreds of Bureh's fighters were killed. Bai Bureh was finally captured on November 11, 1898 and sent into exile in the Gold Coast (now Ghana), while 96 of his comrades were hanged by the British.

The defeat of the natives in the Hut Tax war ended large scale organised resistance to colonialism; however resistance continued throughout the colonial period in the form of intermittent rioting and chaotic labour disturbances. Riots in 1955 and 1956 involved "many tens of thousands" of natives in the protectorate.

One notable event in 1935 was the granting of a monopoly on mineral mining to the Sierra Leone Selection Trust run by De Beers, which was scheduled to last 98 years.

An Independent Nation

The 1924 Sierra Leone constitution was divided into Colony and Protectorate with separate and different political systems. Antagonism between the two entities came to a heated debate in 1947 when proposals were introduced to provide for a single political system for both the Colony and the Protectorate, most of whom were to come from the Protectorate. The Creoles, lead by Isaac Wallace-Johnson naturally opposed the proposals whose effect would have been to diminish their political power. It was due to the astute politics of Sir Milton Margai an ethnic Mende and the leading politician from the protectorate that educated Protectorate elite were won over to join forces with the paramount chiefs in the face of Creole intransigence. Later, Sir Milton use the same skills to win over opposition leaders and moderate Creole elements for the achievement of independence.

In November 1951 Sir Milton Margai oversaw the drafting of a new Sierra Leone constitution which united the formerly separate Colonial and Protectorate legislatures and — most importantly — provided a framework for decolonization. In 1953 Sierra Leone was granted local ministerial powers and Milton Margai was made Chief Minister under the new constitution. The new constitution ensured Sierra Leone a parliamentary system within the Commonwealth of Nations and was formally adopted in 1958.

Margai led the Sierra Leonean delegation at the constitutional conferences that were held with British Colonial Secretary Iain Macleod in London in 1960. Most of Sierra Leone's most prominent politicians were in the delegation, including Sir Milton's younger brother Sir Albert Margai, Sir Banja Tejan Sie, Kande Bureh, John Kareefa Smart, Lamina Sankoh and Ella Koblo Gulama, the only female in the Sierra Leonean delegation. One notable absentee was Siaka Stevens, the leader of the opposition APC who was under house arrest in Freetown after being accused of disrupting the Independence movement. On April 27, 1961, Milton Margai led Sierra Leone to independence from the United Kingdom. Thousands of Sierra Leoneans throughout the nation took to the street to celebrate their Independence. The nation held its first general elections on May 27, 1962 and Margai was elected Sierra Leone's first Prime Minister by a landslide. Milton Margai's political party, the Sierra Leone People's Party (SLPP), won by large margins in the nation's first general election under universal adult suffrage in May 1962. An important aspect of Sir Milton's character was his self-effacement. He was neither corrupt nor did he make a lavish display of his power or status. Sir Milton's government was based on the rule of law and the notion of separation of powers, with multiparty political institutions and fairly vibrant representative structures. Milton Margai used his conservative ideology to lead Sierra Leone without much strife. He appointed government officials with a clear eye to satisfy various ethnic groups.

Upon Margai's death on April 28, 1964, his brother Sir Albert Margai, who was Sierra Leone's Minister of Finance under Milton's government was chosen by an overwhelming majority vote in Parliament to be the new leader of the SLPP and the country's next prime minister. Sir Albert Margai's leadership was briefly challenged by Sierra Leone's Foreign Minister John Karefa-Smart, an ethnic Sherbro, who questioned Sir Albert succession to the SLPP leadership position. Kareefa-Smart received little support in Parliament to have Margai stripped of the SLPP leadership.

Sir Albert was sworn in as Sierra Leone's second Prime Minister the same day his brother died at a ceremony held at the state house in Freetown. Soon after Margai was sworn in as Prime Minister, he immediately dismissed Karefa-Smart and several other senior government officials who had served under his elder brother Sir Milton's government, as he viewed them as traitors and a threat to his administration.

Sir Albert was highly criticized during his three-year rule as prime minister. He was accused of corruption and of favouritism toward his own Mende ethnic group. Albert Margai, unlike his late brother Milton, was opposed to the colonial legacy of allowing the country's Paramount Chiefs executive powers and he was seen as a threat to the existence of the ruling houses across the country. This made him unpopular with the powerful paramount chiefs, most of whom were founding members of the SLPP. To strengthen support for his reform agenda for the party and the country the new Pime Minister brought into the executive of the SLPP and his government younger, western-educated, and more radicallised members of the party including Salia Jusu Sheriff (PhD). The party was thus divided with the traditionalist and more powerful old guard against the new and younger leaders.

Sir Albert also tried to establish a one-party state but with very little support in Parliament, even among his fellow SLPP members and was also met by fierce resistance from the main opposition the All People's Congress (APC), which had become suddenly more popular than the rulling SLPP and ultimately abandoned the idea. During Albert Margai's administration, The Mende increased their influence both in the civil service and the army. Most of the top military and government positions were held by Mendes, and Mende country (the South-Eastern part of Sierra Leone) received preferential treatment.

In closely contested general elections in March 1967, Sierra Leone Governor General Henry Josiah Lightfoot Boston declared the new prime minister to be Siaka Stevens, an ethnic Limba and the candidate of the All People's Congress (APC) and the Mayor of Freetown. Stevens was sworn in as Sierra Leone's third Prime Minister on May 17, 1967 in Freetown. Hours after taking office, Stevens was ousted in a bloodless coup led by the commander of the Sierra Leone Armed Forces, Brigadier David Lansana, an ethnic Mende and a prominent supporter of Albert Margai who had appointed him in 1964 as the commander of the Sierra Leone Armed Forces. Brigadier Lansana insisted that the determination of office should await the election of the tribal representatives in Parliament, mostly from Mende chiefdoms in South-Eastern Sierra Leone.

Stevens was placed under house arrest and martial law was declared. However, on March 23, 1968, a group of senior military officers in the Sierra Leone Army, lead by Brigadier Andrew Juxon-Smith, an ethnic Creole, overrode this action by seizing control of the government, arresting Lansana and suspending the constitution. The group constituted itself as the National Reformation Council (NRC) with Brigadier Andrew Juxon-Smith as its chairman. In April 1968, the NRC was overthrown by a group of senior military officers in the Sierra Leone Army who called themselves the Anti-Corruption Revolutionary Movement (ACRM), led by Brigadier General John Amadu Bangura, an ethnic Limba. The ACRM imprisoned Brigadier Andrew Juxon-Smith and other senior NRC members, restored the constitution and reinstated Stevens as Prime Minister.

Stevens assumed power once again in 1968 with a great deal of promise and ambition. Much trust was placed upon him as he was then champion of multi-party politics. Upon taking power from the military, however, he soon strove to drive the SLPP from competitive politics in various general elections, using violence and intimidation.

After the return to civilian rule, by-elections were held (beginning in autumn 1968) and an all-APC cabinet was appointed. Calm was not completely restored. In November 1968, Stevens declared a state of emergency after disturbance in the provinces.

Stevens had campaigned on a platform of socialist principles. However, when he became Prime Minister he abandoned his pre-election promises and employed an authoritarian model of governance. Many senior officers in the Sierra Leone military were greatly disappointed but none could confront Stevens. Brigadier General Bangura who had reinstated Stevens as Prime Minister was widely considered the only person who could put the brakes on Stevens. Bangura was a magnetic and popular figure among Sierra Leoneans. The army was devoted to him and this made him potentially dangerous to Steven's new agenda in the shifting political climate of Sierra Leone. In January 1970, Bangura was arrested and charged with conspiracy and plotting to commit a coup against the Stevens government. He was convicted and sentenced to death by execution. On March 29, 1970, Stevens had Bangura hanged at the Kissy Road in central Freetown.

In March 1971 the government survived an unsuccessful military coup. The coup leaders were executed, including several senior officers in the army and some senior government officials.

On April 19, 1971, parliament declared Sierra Leone a republic, with Siaka Stevens as President. Guinean troops requested by Stevens to support his government were in the country from 1971 to 1973. In May 1973 a general elections were held throughout the country, but the main opposition the SLPP boycotted the 1973 general election, alleging widespread intimidation and procedural obstruction..

In 1973 president Stevens and president William Tolbert of Liberia signed a treaty forming the Mano River Union to facilitate trade between Sierra Leone and Liberia – with Guinea joining in 1980 under president Sekou Toure. In 1975 Sierra

Leone joint the Economic Community of West African States (commonly known as ECOWAS).

An alleged plot to overthrow president Stevens failed in 1974 and its leaders were executed. In March 1976, Stevens was elected without opposition for a second five-year term as president. On July 19, 1975, 14 senior army and government officials including Brigadier David Lansana, former cabinet minister Mohamed Sorie Forna, Brigadier General Ibrahim Bash Taqi and Lieutenant Habib Lansana Kamara were executed after being convicted for allegedly attempting a coup to topple president Stevens' government.

In early 1977 a major anti-government demonstration by students and youth occurred throughout the country against the APC government and deteriorating economic conditions. But the demonstration was put down by the police and the army.

In the national parliamentary election of May 1977, the APC won 74 seats and the main opposition, the SLPP, won 15. The SLPP condemned the election, alleged widespread vote-rigging and voter intimidation. In 1978, the APC dominant parliament approved a new constitution making the country a one-party state. The 1978 referendum made the APC the only legal political party in Sierra Leone. This move lead to another major demonstration in many parts of the country but again it was put dawn by the army and the police.

Under the APC regimes headed by Stevens, the Limba, Stevens' own ethnic group, enjoyed strong influence in the government and civil service. During the 1970s, another major ethnic group, the Temne joined the Mende in opposition to the APC government. But after Stevens appointed a Temne, Sorie Ibrahim Koroma as vice-president in 1978, the Temne appeared to have emerged as an influential group in the government.

Stevens is generally criticised for dictatorial methods and government corruption, but, on a positive note, he reduced the ethnic polarisation in government by incorporating members of various ethnic groups into his all-dominating APC government. Siaka Stevens retired in November, 1985 after being in power for 18 years, but continued to be chairman of the APC. The APC named a new presidential candidate to succeed Stevens at their last delegate conference held in Freetown in November 1985. He was Major General Joseph Saidu Momoh, the commander of the Republic of Sierra Leone Armed Forces and Stevens' own choice to succeed him. As head of the Sierra Leone Armed Forces, Major General Momoh was very loyal to Stevens who had appointed him to the position. Like Stevens, Momoh was also a member of the minority Limba ethnic group. Joseph Saidu Momoh was elected President in a one-party parliament as the only contesting candidate. Momoh was sworn in as Sierra Leone's second president on November 28, 1985 at an inauguration ceremony held in Freetown. A one party parliamentary elections between APC members were held in May, 1986.

President Momoh's strong links with the army and his verbal attacks on corruption earned him much needed initial support among Sierra Leoneans. With the lack of new faces in the new APC cabinet under president Momoh and the return of many of the old faces from Stevens government, criticisms soon arose that Momoh was simply perpetuating the rule of Stevens. The next couple of years under the Momoh administration were characterised by corruption, which Momoh defused by sacking several senior cabinet ministers. To formalise his war against corruption, President Momoh announced a Code of Conduct for Political Leaders and Public Servants"

After an alleged attempt to overthrow President Momoh in March 1987, more than 60 senior government officials were arrested, including Vice-President Francis Minah, who was removed from office, convicted for plotting the coup, and executed by hanging in 1989 along with 5 others.

Multi-party Constitution and Revolutionary United Front Rebellion

In October 1990, due to mounting pressure from both within and outside the country for political and economic reform, president Momoh set up a constitutional review commission to review the 1978 one-party constitution. Based on the commission's recommendations a constitution re-establishing a multi-party system was approved by the exclusive APC Parliament by a 60% majority vote, becoming effective on

October 1, 1991. By November 1991, political opposition became active once again in Sierra Leone. In late November 1991, president momoh proposed a multi-party presidential and parliamentary elections to be held in the country in October 1992.

There was great suspicion that president Momoh was not serious about his promise of political reform, as APC rule continued to be increasingly marked by abuses of power. The APC was also alleged to have been hording arms and planning a violent campaign against the opposition parties ahead of multi-party general elections scheduled for late 1992. Several senior government officials in the APC administration like Dr. Salia Jusu Sheriff, Dr. Abass Bundu, J.B. Dauda and Dr. Sama Banya resigned from the APC government respectively to resuscitate the previously disbanded SLPP. While other senior government officials like Thaimu Bangura, Edward Kargbo and Desmond Luke resigned from the APC and formed their own respective political parties to challenge the ruling APC.

Civil war broke out, mainly due to government corruption and mismanagement of diamond resources and abuse of power by various governments since independence from Britain (Truth and Reconciliation Commission's Report). The brutal civil war going on in neighbouring Liberia played an undeniable role in the outbreak of fighting in Sierra Leone. Charles Taylor—then leader of the National Patriotic Front of Liberia—reportedly helped form the Revolutionary United Front (RUF) under the command of former Sierra Leonean army corporal Foday Saybana Sankoh, an ethnic Temne from Tonkolili District in Northern Sierra Leone. Sankoh was a British trained former army corporal who had also undergone guerrilla training in Libya. Taylor's aim was for the RUF to attack the bases of Nigerian dominated peacekeeping troops in Freetown who were opposed to his rebel movement in Liberia. In 2003 Foday Sankoh was indicted by the Special Court for Sierra Leone for war crimes and crimes against humanity and died under UN custody before the trials could be concluded. Charles Taylor, who is a former president of Liberia, is currently in the International Criminal Court (ICC) in the Hague facing charges of war crimes committed by Sankoh's RUF in Sierra Leone. Taylor is the first former president to face the ICC. The RUF, led by Sankoh and backed by Taylor, launched its first attack in villages in Kailahun District in Eastern Sierra Leone from Liberia on March 23, 1991. The government of Sierra Leone, overwhelmed by a crumbling economy and corruption, and a demoralised army was unable to put up significant resistance against the incursion of the RUF. Within a month of entering Sierra Leone from Liberia, the RUF controlled much of Eastern Sierra Leone, including the cash crop production areas of Kailahun and the government diamond mines in Kono District. Forced recruitment of child soldiers was also an early feature of the rebel strategy.

On April 29, 1992, a group of six young soldiers in the Sierra Leonean army, apparently frustrated by the government's failure to deal with the rebels and a revolt over salary pay launched a military coup which sent president Momoh into exile in Guinea. The young soldiers were Second Lieutenant Tom Nyuma, an ethnic Kissi, Captain Valentine E. M. Strasser, an ethnic Creole, Captain Solomon A. J. Musa, an ethnic Mende, Captain Samuel Komba Kambo, an ethnic Kono, Captain Julius Maada Bio, an ethnic Mende, Captain Komba Mondeh, an ethnic Kono and Colonel Yahya Kanu, an ethnic Temne. A seventh member of the leadership of the coup, Lieutenant Sahr Sandy (who was said to have insisted to his collegues on the second day of the standoff that theirs would no longer be a revolt over pay but a "Revolution" to overthrow the APC and the system of One Party Politics), was killed during the coup, allegedly by his adopted uncle and member of the APC Colonel S.I.M.

Turay (Turay was declared wanted for the murder of Sandy but managed to make it across to Guinea where he joined the exiled president. He returned to Freetown following the 2007 election that brought APC back to power, and he has not been cleared/arrested for Sandy's death). According to many insiders including Tom Nyuma, had he been alive, Sahr Sandy would have been the head of state following the arrest of Colonel Yahya Kanu. Colonel Yahya Kanu was the very popular commander of the fearless Tiger Battalion which was at the forefront in the war against the RUF under president Momoh. The soldiers established the National Provisional Ruling Council

(NPRC). Though he was not formally declared head of the new junta, and even refused to accept in an interview with the BBC Focus on Africa Program that the revolt was a coup by his men, Colonel Kanu was seen as the defacto leader of the NPRC. Kanu was later arrested and imprisoned by his junior officers, who accused him of trying to negotiate a compromise with the toppled APC administration. Kanu's arrest divided the army into two rival groups, namely, his Tiger Battalion and Tom Nyuma's Cobra Battalion and their respective supporters.

On April 29, 1992, 27-year-old Valentine Strasser took over as leader and chairman of the NPRC and Head of State of Sierra Leone. Strasser became the youngest Head of State in the world, just three days after his 27 birthday. 25 year-old S.A.J. Musa, a close friend of Strasser and an officer in Kanu's feared Tiger Battalion, was named Vice-Chairman of the NPRC. Many Sierra Leoneans nationwide rushed into the streets to celebrate the NPRC's takeover from the 23-year dictatorial APC regime, which they perceived as corrupt.

The NPRC junta immediately suspended the 1991 Constitution, declared a state of emergency, banned all political parties, limited freedom of speech and freedom of the press, and enacted a rule-by-decree policy, in which soldiers were granted unlimited powers of administrative detention without charge or trial, and challenges against such detentions in court were precluded. The NPRC Junta maintained relations with ECOWAS and strengthened support for ECOMOG troops based in Sierra Leone which was fighting in Liberia. In his first speech as head of state the young Strasser reassured the world of meeting his country's obligations to her creditors, and making a commitment to the IMF and the World Bank to accelerate the economic reform process started by Momoh's government in 1989 aimed at stabilizing the severely crippled economy and shortly after that went on to negotiate a Structural Adjustment Program with these two institutions. The junta formed the Supreme Council of State (SCS) made up of only the leading figures of the NPRC including the six ring leaders and chaired by Strasser himself. They also appointed an advisory council which consisted of retired senior civil servants and academics, chaired by a retired UN administrator Ahmad Tejan Kabbah, who were all regarded as untainted by the 23 years of alleged APC mismanagement, curruption and abuse of power.

In December 1992, an alleged coup attempt to topple the NPRC administration of Strasser and release the detained Colonel Yayah Kanu was foiled. Sgt. Lamin Bangura (an ethnic Temne) and some junior army officers of the Tiger Battalion were identified to be behind the alleged coup plot. It led to the execution of seventeen soldiers in the Sierra Leone Army, including Sgt.

Bangura and Yayah Kanu, and some senior members of the overthrown APC government who had been in detention at the Pa Demba Road prison including the notorious Inspector General of Police James Bambay Kamara, key former APC ministers, senior party members and thugs. By mid 1993 captain Strasser announced a plan to hand over the government to civilian rule by 1996. Dr. James Jonah, who was by then Deputy Secretary General of the United Nations, was appointed by the NPRC Junta as the chairman of the new Interim National Electoral Commission (INEC) which was in charge of the demarcation of electoral boundaries and voters registration. In 1994 the NPRC junta proposed a change in the age restriction in the 1991 Sierra Leone constitution which stated only Sierra Leoneans over the age of 40 are eligible for the presidency, thus excluding Strasser and others in the NPRC.

The NPRC proved to be nearly as ineffectual as the Momoh-led APC government in repelling the RUF. More and more of the country fell to RUF fighters, and by 1995 they held much of the diamond-rich Eastern Province and were at the edge of Freetown. In response, the NPRC hired several hundred mercenaries from the private firm Executive Outcomes. Within a month they had driven RUF fighters back to enclaves along Sierra Leone's borders, and cleared the RUF from the Kono diamond producing areas of Sierra Leone.

However, Captains Tom Nyuma (Secretary of State East) and Komba Mondeh (Secretary of State Defence), who were regarded outside Freetown as the only "fighters" in the NPRC who dared to lead the troops to attack RUF strongholds in the East and South, were widely credited for these successes against the RUF. During this time corruption had erupted within the senior ranks of both the NPRC and the military, and the junta had become divided between SAJ Musa, on the one side, against Nyuma and Mondeh, on the other.

SAJ Musa had become very popular in Freetown for fighting grafts and enforcing strict discipline in the public service and his last-Saturday-of-the-month city cleaning exercises. Nyuma, nick named "The Ranger", was seen across the country as the daredevil of the NPRC and the "protector of the East". There was great suspicion among the SCS members that SAJ Musa was planning a coup to topple his friend Strasser, whom he accused of being subservient to the wishes of Nyuma and Mondeh. On July 5, 1995, under pressure from Nyuma, Captain Strasser dismissed SAJ Musa as deputy chairman of the NPRC and appointed an ally of Tom Nyuma, the Secretary of State for Information and Broadcasting Captain Julius Maada Bio, to the position. Musa was arrested by a group of soldiers led by Nyuma's men, and was briefly placed under house arrest in Freetown before he was sent on exile to the UK. Senior NPRC members, including Bio (who by now had promoted himself to Brigadier), Nyuma and Mondeh (both promoted to Colonel), were becoming increasingly unhappy with Strasser's (who remained Captain) handling of the preparation for the pending elections, the peace negotiation with the RUF, and the transition to democratic civilian rule.

In January 1996, after nearly four years in power, Captain V.E.M Strasser was ousted in a bloodless "palace" coup led by his NPRC deputy Brigadier General Julius Maada Bio with the support of several senior NPRC members including both Tom Nyuma and Komba Mondeh. Bio claimed that Strasser was attempting to unilaterally amend the age restriction aspect of the constitution in order to perpetuate his hold on power. Bio reinstated the Constitution and called for general elections. In the second round of presidential elections in early 1996, Ahmad Tejan Kabbah, and ethnic Mandingo and the candidate of the Sierra Leone People's Party (SLPP), won 59% of the vote, over John Karefa-Smart, an ethnic Sherbro and the candidate of the United National People's Party (UNPP) who won 41%. Bio fulfilled his promise of a return to civilian rule, and handed power to Kabbah. President Tejan Kabbah's SLPP party also won a majority of the seats in Parliament. For years Sierra Leonean Soldiers in the lower ranks were not paid a good salary and they were denied privileges and benefits. Soldiers were killed at the war front and no provision was made for their families. Major Johnny Paul Koroma, an officer in the Sierra Leone army who hail from the Limba ethnic group from Bombali District was allegedly involved in an attempt to overthrow the government of president Kabbah. He was arrested, tried, convicted, and imprisoned at Freetown's Pademba Road Prison.

On May 25, 1997, a group of seventeen junior officers in the Sierra Leone army, loyal to Major Koroma, who called themselves the Armed Forces Revolutionary Council (AFRC) led by Corporal Tamba Gborie and Sergent Alex Tamba Brima, both ethnic Kono, launched a military coup which sent president Kabbah into exile in Guinea. Corporal Tamba Gborie quickly went to the SLBS FM 99.9 headquarter in Freetown to announce the coup to a shocked nation, and to alert all soldiers throughout the nation on guard duty. The AFRC released Koroma from prison and installed him as their chairman and Head of State of the country with Corporal Tamba Gborie as deputy in command of the AFRC.

Koroma suspended the constitution, banned demonstrations, shut down all private radio stations in the country and invited the RUF to join the new junta government, with its leader Foday Sankoh as the Vice-Chairman of the new AFRC-RUF coalition junta government. Within days Freetown was overwhelmed by the presence of the RUF combatants who came to the city in their thousands. The Kamajors, a group of traditional fighters mostly from the Mende ethnic group under the command of deputy Defence Minister Samuel Hinga Norman remained loyal to President Kabbah. The Kamajors defended Bo, the country's second largest city from the Junter and continue their attack against the AFRC and RUF in south-eastern Sierra Leone

After 10 months in office, the junta was ousted by the Nigeria-led ECOMOG forces, and the democratically elected government of president Kabbah was reinstated in March 1998. Hundreds of civilians who had been accused of helping the

AFRC government were illegally detained. Courts-martial were held for soldiers accused of assisting the AFRC government. Twenty-four of these were found guilty and were executed without appeal in October 1998. On January 6, 1999, AFRC made another unsuccessful attempt to overthrow the government, causing an estimated 3,000 deaths, raping women and girls, abdduction and subsequent conscription of children, amputations of limbs, and much destruction of property in and around Freetown.

In October, the United Nations agreed to send peacekeepers to help restore order and disarm the rebels. The first of the 6,000-member force began arriving in December, and the UN Security Council voted in February 2000 to increase the force to 11,000, and later to 13,000. But in May, when nearly all Nigerian forces had left and UN forces were trying to disarm the RUF in eastern Sierra Leone, Sankoh's forces clashed with the UN troops, and some 500 peacekeepers were taken hostage as the peace accord effectively collapsed. The hostage crisis resulted in more fighting between the RUF and the government.

The situation in the country deteriorated to such an extent that British troops were deployed in Operation Palliser, originally simply to evacuate foreign nationals. However, the British exceeded their original mandate, and took full military action to finally defeat the rebels and finally restore order, and the British were the catalyst for the ceasefire that ended the civil war.

Elements of the British Army, together with administrators and politicians remain in Sierra Leone to this day, helping train the armed forces, improve the infrastructure of the country and administer financial and material aid. Tony Blair, the Prime Minister of Britain at the time of the British intervention is regarded as a hero by the people of Sierra Leone, many of whom are keen for more British involvement.

Between 1991 and 2001, about 50,000 people were killed in Sierra Leone's civil war. Hundreds of thousands of people were forced from their homes, and many became refugees in Guinea and Liberia. In 2001, UN forces moved into rebel-held areas and began to disarm rebel soldiers. By January 2002, the war was declared over. In May, Kabbah was reelected president.

By 2004, the disarmament process was complete. Also in 2004, a UN-backed war crimes court began holding trials of senior leaders from both sides of the war. In December 2005, UN peacekeeping forces pulled out of Sierra Leone.

In August 2007, Sierra Leone held presidential and parliamentary elections. However, no presidential candidate won the 50% plus one votes majority stipulated in the constitution on the first round of voting. A runoff election was held in September 2007, and Ernest Bai Koroma, the candidate of the APC and ethnically a half Limba and half Temne from the north was elected president.

By 2007, there had been an increase in the number of drug cartels, many from Colombia, using Sierra Leone as a base to ship drugs on to Europe. It was feared that this might lead to increased corruption and violence and turn the country, like neighbouring Guinea-Bissau, into a narco state. However, the new government of president Koroma quickly amended the laws against drug trafficking in the country, updating the existing legislation from those inhereted at independence in 1961, to address the international concerns, increasing punishment for offenders both in terms of higher, if not prohibitive, fines, lengthier prison terms and provision for possible extradition of offenders wanted elsewhere including to the United States of America.

Geography and Climate

Sierra Leone is located on the west coast of Africa, between the 7th and 10th parallels north of the equator. Sierra Leone is bordered by Guinea to the north and northeast, Liberia to the south and southeast, and the Atlantic Ocean to the west.

The country has a total area of 71,740 km (27,699 sq mi), divided into a land area of 71,620 km (27,653 sq mi) and water of 120 km (46 sq mi). The country has four distinct geographical regions. In eastern Sierra Leone the plateau is interspersed with high mountains, where Mount Bintumani reaches 1,948 m (6,391 ft), the highest point in the country. The upper part of the drainage basin of the Moa River is located in the south of this region. The centre of the country is a region of lowland plains, containing forests, bush and farmland, that occupies about 43% of Sierra Leone's land area. The northern section of this has been categorised by the World Wildlife Fund as part of the Guinean forest-savanna mosaic ecoregion, while the south is rain-forested plains and farmland. In the west Sierra Leone has some 400 km (249 mi) of Atlantic coastline, giving it both bountiful marine resources and attractive tourist potential. The coast has areas of low-lying Guinean mangroves swamp. The national capital Freetown sits on a coastal peninsula, situated next to the Sierra Leone Harbour, the world's third largest natural harbour.

Climate

The climate is tropical, with two seasons determining the agricultural cycle: the rainy season from May to November, and a dry season from December to May, which includes harmattan, when cool, dry winds blcw in off the Sahara Desert and the night-time temperature can be as low as 16 °C (60.8 °F). The average temperature is 26 °C (78.8 °F) and varies from around 26 °C (78.8 °F) to 36 °C (96.8 °F) during the year.

Environment

Logging, mining, slash and burn, and deforestation for land conversion-such as cattle grazing-have dramatically diminished forested land in Sierra Leone since the 1980s. Correspondingly the habitat for the African Wild Dog, *Lycaon pictus*, has been decreased, such that this canid is deemed to have been extirpated in Sierra Leone.

Until 2002, Sierra Leone lacked a forest management system due to a brutal civil war that caused tens of thousands of deaths. Deforestation rates have increased 7.3% since the end of the civil war. On paper, 55 protected areas covered 4.5% of Sierra Leone as of 2003. The country has 2,090 known species of higher plants, 147 mammals, 626 birds, 67 reptiles, 35 amphibians, and 99 fish species.

The Environmental Justice Foundation has documented how the number of illegal fishing vessels in Sierra Leone's waters has multiplied in recent years. The amount of illegal fishing has significantly depleted fish stocks, depriving local fishing communities of an important resource for survival. The situation is particularly serious as fishing provides the only source of income for many communities in a country still recovering from over a decade of civil war.

In June 2005, the Royal Society for the Protection of Birds (RSPB) and Bird Life International agreed to support a conservation-sustainable development project in the Gola Forest in southeastern Sierra Leone, an important surviving fragment of rainforest in Sierra Leone.

Government and Politics

Sierra Leone is a constitutional republic with a directly elected president and a unicameral legislature. The current system of government in Sierra Leone, established under the 1991 Constitution, is modelled on the following structure of government: the Legislature, the Executive and the Judiciary.

Within the confines of the 1991 Constitution, supreme legislative powers are vested in Parliament, which is the law making body of the nation. Supreme executive authority rests in the president and members of his cabinet and judicial power with the judiciary of which the Chief Justice is head.

The president is the head of state, the head of government and the commander-in-chief of the Sierra Leone Armed Forces and the Sierra Leone Police. The president appoints and heads a cabinet of ministers, which must be approved by the Parliament. The president is elected by popular vote to a maximum of two five-year terms. The president is the highest and most influential position within the government of Sierra Leone.

To be elected president of Sierra Leone, a candidate must gain at least 55% of the vote. If no candidate gets 55%, there is to be a second-round runoff between the top two candidates.

The current president of Sierra Leone is Ernest Bai Koroma, who was sworn in on September 17, 2007, shortly after being declared the winner of a tense run-off election over the incumbent Vice president, Solomon Berewa of the Sierra Leone People's Party (SLPP).

Next to the president is the Vice president, who is the second-highest ranking government official in the executive branch of the Sierra Leone Government. As designated by the

Sierra Leone Constitution, the vice president is to become the new president of Sierra Leone upon the death, resignation, or removal of the president by parliament and to assume the Presidency temporarily while the president is otherwise temporarily unable to fulfill his or her duties. The vice president is elected jointly with the president as his or her running mate. Sierra Leone's current vice president is Samuel Sam-Sumana, sworn in on September 17, 2007.

The Parliament of Sierra Leone is unicameral, with 124 seats. Each of the country's fourteen districts is represented in parliament. 112 members are elected concurrently with the presidential elections; the other 12 seats are filled by paramount chiefs from each of the country's 12 administrative districts.

The current parliament in the August 2007 Parliamentary elections is made up of three political parties. The most recent parliamentary elections were held on August 11, 2007. The All People's Congress (APC), won 59 of 112 parliamentary seats; the Sierra Leone People's Party (SLPP) won 43; and the People's Movement for Democratic Change (PMDC) won 10. To be qualified as Member of Parliament, the person must be a citizen of Sierra Leone, must be at least 21 years old, must be able to speak, read and write the English language with a degree of proficiency to enable him to actively take part in proceedings in Parliament; and must not have any criminal conviction.

Since independence in 1961, Sierra Leone's politics has been dominated by two major political parties, the Sierra Leone People's Party (SLPP), and the ruling All People's Congress (APC), although other minor political parties have also existed but with no significant supports.

The judicial power of Sierra Leone is vested in the judiciary, headed by the Chief Justice and comprising the Sierra Leone Supreme Court, which is the highest court in the country and its ruling therefore cannot be appealed; High Court of Justice; the Court of Appeal; the magistrate courts; and traditional courts in rural villages. The president appoints and parliament approves Justices for the three courts. The Judiciary have jurisdiction in all civil and criminal matters throughout the country. The current Sierra Leone's Chief Justice is Umu Hawa

Tejan Jalloh, who was appointed by President Ernest Bai Koroma and took office on January 25, 2008 upon her confirmation by parliament. She is the first woman in the history of Sierra Leone to hold such position.

Foreign Relations

The Sierra Leone Ministry of Foreign Affairs and International Relations, headed by Minister of Foreign Affairs Zainab Hawa Bangura is responsible for foreign policy of Sierra Leone. Sierra Leone has diplomatic relations that include China, Libya, Iran, and Cuba. Sierra Leone has good relations with the West, including the United States and has maintained historical ties with the United Kingdom and other former British colonies through membership of the Commonwealth of Nations. Great Britain has played a major role in providing aid, together with administrative help and military training since intervening to end the Civil War in 2000.

Former President Siaka Stevens' government had sought closer relations with other West African countries under the Economic Community of West African States (ECOWAS) a policy continued by the current. Sierra Leone, along with Liberia and Guinea form the Mano River Union (MRU) primarily designed to implement development projects and promote regional economic integration between the three countries.

Sierra Leone is also a member of the United Nations and its specialized agencies, the African Union, the African Development Bank (AFDB), the Organization of the Islamic Conference (OIC), and the Non-Aligned Movement (NAM). Sierra Leone is also a member of the International Criminal Court with a Bilateral Immunity Agreement of protection for the US military (as covered under Article 98).

Provinces and Districts

The Republic of Sierra Leone is composed of three provinces: the Northern Province, Southern Province and the Eastern Province and one other region called the Western Area. The provinces are further divided into 12 districts, and the districts are further divided into chiefdoms, except for the Western Area.

Major Cities

City	2004 Census	Current Population Estimate
Freetown	772,873	1,070,200
Bo	149,957	215,474
Kenema	128,402	164,125
Koidu Town	80,025	111,800
Makeni	82,840	105,900

The populations quoted above for the five largest cities are estimates from the sources cited. Different sources give different estimates. Some claim that Magburaka should be included in the above list, but one source estimates the population at only 14,915, whilst another puts it as high as 85,313.

Economy

Sierra Leone is slowly emerging from a protracted civil war and is showing signs of a successful transition. Investor and consumer confidence continue to rise, adding impetus to the country's economic recovery. There is greater freedom of movement and the successful re-habitation and resettlement of residential areas.

Rich in minerals, Sierra Leone has relied on mining, especially diamonds, for its economic base. The country is among the top 10 diamond producing nations in the world. Mineral exports remain the main foreign currency earner. Sierra Leone is a major producer of gem-quality diamonds. Though rich in diamonds, it has historically struggled to manage their exploitation and export.

Annual production of Sierra Leone's diamond estimates range between $250–300 million U.S dollar. Some of that is smuggled, where it is possibly used for money laundering or financing illicit activities. Formal exports have dramatically improved since the civil war with efforts to improve the management of them having some success. In October 2000, a UN-approved certification system for exporting diamonds from the country was put in place and led to a dramatic increase in legal exports. In 2001, the government created a mining community development fund, which returns a portion of diamond export taxes to diamond mining communities. The fund was created to raise local communities' stake in the legal diamond trade.

Sierra Leone is perhaps best known for its blood diamonds that were mined and sold to De Beers and other diamond conglomerates during the civil war, and whose monies were used to buy the weapons that fueled the atrocities of the civil war. In the 1970s and early 1980s, economic growth rate slowed because of a decline in the mining sector and increasing corruption among government officials.

By the 1990s economic activity was declining and economic infrastructure had become seriously degraded. Over the next decade much of the formal economy was destroyed in the country's civil war. Since the end of hostilities in January 2002, massive infusions of outside assistance have helped Sierra Leone begin to recover. Much of the recovery will depend on the success of the government's efforts to limit corruption by officials, which many feel was the chief cause for the civil war. A key indicator of success will be the effectiveness of government management of its diamond sector.

Sierra Leone has one of the world's largest deposits of rutile, a titanium ore used as paint pigment and welding rod coatings. Sierra Rutile Limited, owned by a consortium of United States and European investors, began commercial mining operations near the city of Bonthe, in the Southern Province, in early 1979. It was then the largest non-petroleum US investment in West Africa. The export of 88,000 tons realized $75 million in export earnings in 1990. In 1990, the company and the government made a new agreement on the terms of the company's concession in Sierra Leone. Rutile and bauxite mining operations were suspended when rebels invaded the mining sites in 1995, but exports resumed in 2005.

About two-thirds of the population engages in subsistence agriculture, which accounts for 52.5% of national income. The government is trying to increase food and cash crop production and upgrade small farmer skills. The government works with several foreign donors to operate integrated rural development and agricultural projects.

Despite its successes and development, the Sierra Leone economy still faces significant challenges. There is high unemployment, particularly among the youth and ex-combatants. Authorities have been slow to implement reforms in the civil service, and the pace of the privatisation programme is also slacking and donors have urged its advancement.

Currency

Sierra Leone's currency is the Leone. The central bank of the country is the Bank of Sierra Leone which is located in the capital, Freetown.

Sierra Leone operates a floating exchange rate system, and foreign currencies can be exchanged at any of the commercial banks, recognised foreign exchange bureaux and most hotels.

Credit card use is limited in Sierra Leone, though they may be used at some hotels and restaurants. There are a few internationally linked automated teller machines that accept Visa cards in Freetown operated by Pro Credit Bank.

Demographics

The 2009 UN estimate of Sierra Leone's population is 6.4 million. Freetown, with an estimated population of 1,070,200, is the capital, largest city, and the hub of the economy, commercial, educational and cultural centre of the country. Bo is the second city with an estimated population of up to 269,000 (149,957 in the 2004 census). Other cities with an estimated population over 100,000 are Kenema, Koidu Town, and Makeni.

Although English is the official language spoken at schools, government administration, and by the media, Krio (language derived from English and several African languages and native to the Sierra Leone Krio people) is the most widely spoken language in virtually all parts of Sierra Leone. The Krio language is spoken by 97% of the country's population and unites all the different ethnic groups, especially in their trade and interaction with each other. In December 2002, Sierra Leone's President Ahmad Tejan Kabbah named Bengali as an "official language" in recognition of the work of 5,300 troops from Bangladesh in the peace-keeping force.

According to the *World Refugee Survey 2008*, published by the U.S. Committee for Refugees and Immigrants, Sierra Leone had a population of 8,700 refugees and asylum seekers at the end of 2007. Nearly 20,000 Liberian refugees voluntarily returned to Liberia over the course of 2007. Of the refugees remaining in Sierra Leone, nearly all were Liberian.

The life expectancy of Sierra Leone is 41 years.

Religion

Religions in Sierra Leone

Religion	***Percent***
Islam	60%
Christianity	30%
Indigenous religions	10%

Followers of Islam are estimated to comprise 60% of Sierra Leone's population. Muslims predominate in all of the country's three provinces and plus the Western Area, though formerly they were concentrated in the north, with the south being mainly Christian. Followers of Christianity comprise about 30% of the population, and those following only African indigenous religion about 10%, though many inhabitants combine traditional beliefs with one of the newer faiths. There are small numbers of adherents to other faiths such as Bahai, Hinduism, and Judaism..

The Sierra Leone constitution provides freedom of religion and the government generally protects this right and does not tolerate its abuse. Unlike many other African countries, the religious diversity of Sierra Leone has seldom led to conflict.

Ethnic Groups

The Sierra Leone government officially recognizes fifteen ethnic groups, each with its own language and custom. Unlike most African nations, Sierra Leone has no serious ethnic divisions and no serious religious divisions. People often married across tribal and religious boundaries.

The two largest and most dominant are the Mende and Temne, each comprises 30% of the population (about 1,888,000 members each). The Mende predominate in the South-Eastern Provinces; the Temne likewise predominate in the Northern Province and the Western Area. Sierra Leone's national politics centres on the competition between the north, dominated by the Temne and the south-east dominated by the Mende. The third largest ethnic group are the Limba at (8.5%) of the population and they are a close ally of the Temne. The Limba are primarily found in Northern Sierra Leone, particularly in Bombali District and they are one of the earliest inhabitant of the country. Sierra Leone's first president Siaka Stevens and the country's second president Joseph Saidu Momoh are ethnic Limba.

The fourth largest ethnic group are the Kono people at around (8%) of the population and they are primarily found in Kono District in Eastern Sierra Leone. The Kono are primarily diamond miners and farmers. Several notable ethnic Kono include the current vice president of Sierra Leone Alhaji Samuel Sam-Sumana and the country's current first lady Sia Koroma, the daughter of former Sierra Leone Attorney General Abu Aiah Koroma.

After the Kono, are the Mandingo (also known as Mandinka) at (7.4%) (they are the descendants of the Mandinka traders from Guinea who immigrated to Sierra Leone between 1840 to about 1898). The Mandinka predominantly fround in the east and the northern part of the country, and they are the largest inhabitant of the large towns, most notably Kabal and Falaba in [[Koinadugu District in the north and Yengema, Kono District in the east of the country. Historically, the Mandinka intense rivals have been the Temne and Limba. Sierra Leone's third president Ahmad Tejan Kabbah is an ethnic Mandingo.

After the Mandinka, are the Fula at (6%) (descendants of 17th-and 18th-century Fulani settlers from the Fouta Djalon region of Guinea) they live primarily in the north and the Western Area of Sierra Leone. Some notable ethnic Fula include the country current chief justice Umu Hawa Tejan Jalloh

Behind the Fula, are the Creole (at 5%) (descendants of freed West Indians slaves from the West Indies and freed African American slaves from the United States which landed in Freetown between 1787 and about 1885) are primarily found in the capital city of Freetown and its surrounding Western Area. Creole culture is unlike that of all other ethnic groups in Sierra Leone, and it is typical of Western culture and ideals. Former Sierrer Leone's Head of State Valentine Strasser is an ethnic Creole. Much smaller ethnic groups are the Kuranko, in the north; the Loko in the north, with the Susu and Yalunka in the far north in Kambia District around the border with Guinea. The Kissi) and the much smaller group of Vai are further inland in Kailahun District in the East next to the border with Liberia. On the coast in Bonthe District in the south are the Sherbro.

In the past, Sierra Leoneans were noted for their educational achievements, trading activity, entrepreneurial skills, and arts and crafts work, particularly wood carving. Many are part of larger ethnic networks extending into several countries, which link West African states in the area. But the level of education and infrastructure has declined sharply over the last 30 years.

Education

Education in Sierra Leone is legally required for all children for six years at primary level (Class P1-P6) and three years in junior secondary education, but a shortage of schools and teachers has made implementation impossible. Two thirds of the adult population of the country are illiterate. The Sierra Leone Civil War resulted in the destruction of 1,270 primary schools and in 2001 67 percent of all school-age children were out of school. The situation has improved considerably since then with primary school enrollment doubling between 2001 and 2005 and the reconstruction of many schools since the end of the war. Students at primary schools are usually 6 to 12 years old, and in secondary schools 13 to 18. Primary education is free and compulsory in government-sponsored public schools.

The country has three universities, the University of Sierra Leone, founded as Fourah Bay College in 1827 (the oldest university in West Africa), and Njala University, primarily located in Bo District, which was established as the Njala Agricultural Experimental Station in 1910 and became a university in 2005. Teacher training colleges and religious seminaries are found in many parts of the country.

Health

Health-care is provided by the government, among others. All medical care is generally charged for in Sierra Leone The country has a very high infant mortality and a very low life expectancy. The maternal death rates are also the highest in the world, at 2,000 deaths per 100,000 live births. The country suffers from epidemic outbreaks of diseases including yellow fever, cholera, lassa fever and meningitis. The prevalence of HIV/AIDS in the population is 1.6 percent, higher than the world average of 1 percent but lower than the average of 6.1 percent in Sub-Saharan Africa.

Military

The Military of Sierra Leone, officially the Republic of Sierra Leone Armed Forces (RSLAF), are the unified armed forces of Sierra Leone responsible for the territorial security of Sierra Leone's border and defending the national interests of Sierra Leone within the framework of its international obligations. The armed forces were formed after independence in 1961, on the basis of elements of the former British Royal West African Frontier Force present in the country. The Sierra Leone Armed Forces currently consist of around 15,500 personnel, comprising the largest Sierra Leone Army, the Sierra Leone Navy and the Sierra Leone Air Wing. The president of Sierra Leone is the Commander in Chief of the military, with the Minister of Defence responsible for defence policy and the formulation of the armed forces. The current Sierra Leone Defence Minister is Ret. Major Alfred Paolo Conteh. The Military of Sierra Leone also has a Chief of the Defence Staff who is a uniformed military official responsible for the administration and the operational control of the Sierra Leone military. Brigadier General Alfred Nelson-Williams who was appointed by president Koroma succeeded the retired Major General Edward Sam M'boma on 12 September 2008 as the Chief of Defence Staff of the Military.

Before Sierra Leone gained independence in 1961 the military was known as the Royal Sierra Leone Military Force. The military seized control in 1968, bringing the National Reformation Council into power. On 19 April 1971, when Sierra Leone became a republic, the Royal Sierra Leone Military Forces were renamed the Republic of Sierra Leone Military Force (RSLMF). The RSLMF remained a single service organization until 1979, when the Sierra Leone Navy was established. It then remained largely unchanged for 16 years until in 1995 when Defence Headquarters was established and the Sierra Leone Air Wing formed. This gave the need for the RSLMF to be renamed the Armed Forces of the Republic of Sierra Leone (AFRSL).

Law Enforcement

Law enforcement in Sierra Leone is primarily the responsibility of the Sierra Leone Police (SLP). Sierra Leone Police was established by the British colony back in 1894 and is one of the oldest police forces in West Africa. The key mission of the Sierra Leone Police include to prevent crime, to protect life and property, to detect and prosecute offenders, to maintain public order, to ensure safety and security, to enhance access to justice. The Sierra Leone Police is headed by the Inspector General of Police, the professional head of the Sierra Leone Police force and is appointed by the President of Sierra Leone. Each one of Sierra Leone's 14 districts is headed by a District Police commissioner who is the professional head of their respective district. The Districts Police Commissioners report directly to the Inspector General of Police at the Sierra Leone Police headquarters in Freetown. The current Inspector General of Police is Brima Acha Kamara who was appointed to the position by former president Ahmad Tejan Kabbah.

Media

Media in Sierra Leone began with the introduction of the first printing press in Africa at the start of the nineteenth century. A strong journalistic tradition developed with the creation of a number of newspapers. In the 1860s, the country became a journalist hub for Africa, with professionals travelling to the country from across the continent. At the end of the nineteenth century, the industry went into decline, and when radio was introduced in the 1930s, it became the primary communication media in the country. The Sierra Leone Broadcasting Service (SLBS) was created by the government in 1934 making it the earliest English language radio broadcaster service in West Africa. The service began broadcasting television in 1963, with coverage extended to all the districts in the country in 1978.

Print media is not widely read in Sierra Leone, especially outside Freetown, partially due to the low levels of literacy in the country. In 2007 there were 15 daily newspapers in the country, as well as those published weekly. Among newspaper readership, young people are likely to read newspapers weekly and older people daily. The majority of newspapers are privately run and are often critical of the government. The standard of print journalism tends to be low due to lack of training, and people trust the information published in newspapers less than that found on the radio.

Radio is the most-popular and most-trusted media in Sierra Leone, with 85% of people having access to a radio and 72% of people in the country listening to the radio daily. These levels do vary between areas of the country, with the Western Area having the highest levels and Kailahun the lowest. Stations mainly consist of local commercial stations with a limited broadcast range, combined with a few stations with national coverage. The United Nations Mission in Sierra Leone (UNIOSIL) runs one of the most popular stations in the country, broadcasting programs in a range of languages. Content includes news of UN activities and human rights information, as well as music and news. The UN missions will withdraw in 2008 and the UN Radio's future is uncertain. There is also a government station run by the SLBS that transmits on FM and short-wave. FM relays of BBC World Service, Radio France Internationale and Voice of America are also broadcast.

Outside the capital Freetown television is not watched by a great many people. There are two national, free terrestrial television stations in Sierra Leone, one run by the government SLBS and the other a private station, ABC Television-Africa (ABC). In 2007, a pay-per-view service was also introduced by GTV as part of a pan-African television service in addition to the nine year old sub-saharan Digital satellite television service (DStv) originating from Multichoice Africa in South Africa. Internet access in Sierra Leone has been sparse but is on the increase, especially since the introduction of wireless services across the country. There are nine Internet Service Providers (ISPs) operating in the country. Freetown has a city wide wireless network and Internet cafes and other businesses offering internet access. Problems experienced with access to the Internet include an intermittent electricity supply and a slow connection speed in the country outside Freetown.

The Sierra Leone constitution guarantees freedom of speech, and freedom of the press; however, the government maintains strong control of media, and at times restricts these rights in practice. Some subjects are seen as taboo by society and members of the political elite; imprisonment and violence have been used by the political establishment against journalists. Under legislation enacted in 1980, all newspapers must register with the Ministry of Information and pay sizable registration fees. The Criminal Libel Law, including Seditious Libel Law of 1965, is used to control what is published in the media. In 2006, President Ahmad Tejan Kabbah committed to reforming the laws governing the press and media to create a freer system for journalists to work in, but in 2007, Sierra Leone was ranked as having the 121st least-free press in the world, with the press less-free, in comparison to other countries, than in 2006.

Transportation

There are a number of systems of transport in Sierra Leone, which has a road, air and water infrastructure, including a network of highways and several airports.

Air

There are ten regional airports in Sierra Leone, and one international airport. The Lungi International Airport located in the coastal town of Lungi in Northern Sierra Leone is the primary airport for domestic and international travel to or from Sierra Leone. Passengers cross the river to Aberdeen Heliports in Freetown by hovercraft, ferry or a helicopter. Helicopters are also available from the airport to other major cities in the country. The airport has paved runways longer than 3,047m. The other airports have unpaved runways, and seven have runways 914 to 1,523 metres long; the remaining two have shorter runways.

Prohibition from E.U. Air Operations

This country appears on the E.U. list of prohibited countries with regard to the certification of airlines. This means that no airline which is Sierra Leone registered may operate services of any kind within the European Union. This is due to substandard safety standards.

Water

Sierra Leone has the third largest natural harbour in the world, where international shipping berth at the Queen Elizabeth II Quay in Government Wharf in central Freetown. There are 800 km of waterways in Sierra Leone, of which 600 km are navigable year-round. Major port cities are Bonthe, Freetown, Sherbro Island and Pepel.

Highways

There are 11,700 kilometres (about 7,270 Miles) of highways in Sierra Leone, of which 936 km (about 581 miles) are paved (about 8% of the roads). Sierra Leone highways are linked to Conakry, Guinea, and Monrovia, Liberia.

Sports

Football

Football (soccer) is by far the most popular sport in Sierra Leone. The national football team, popularly known as the Leone Stars, represents the country in international competitions. It has never qualified for the FIFA World Cup but participated in the 1994 and 1996 African Cup of Nations. The country's national television network, The Sierra Leone Broadcasting Service (SLBS) broadcasts the live match, along with several radio stations throughout the country. Some well known Sierra Leonean footballers include the team captain Mohamed Kallon, Julius Gibrilla Woobay, Al Bangura, Paul Kpaka, Rodney Strasser, Ahmed Deen, Samuel Barlay, Kewullay Conteh Albert Jarrett and Kei Kamara

The Sierra Leone National Premier League is the top football league, controlled by the Sierra Leone Football Association. The two biggest and most successful football clubs are East End Lions and Mighty Blackpool, but Kallon F.C. has enjoyed contemporary success. Kallon F.C. won the Premier League and the Sierra Leonean FA Cup in 2006, and eliminated 2006 Nigerian Premier League Champions Ocean Boys FC in the

2007 CAF Champions League first qualifying round, but later lost to ASEC Mimosas of Ivory Coast in the second qualifying round for the group stage.

The Sierra Leone U-17 football team, nicknamed the Sierra Stars, finished as runner-up at the 2003 African U-17 Championship in Swaziland, but came in last place in their group at the 2003 FIFA U-17 World Championship in Finland.

Cricket

The Sierra Leone cricket team represents Sierra Leone in international cricket competitions, and is among the best in West Africa. It became an affiliate member of the International Cricket Council in 2002. It made its international debut at the 2004 African Affiliates Championship, where it finished last of eight teams. But at the equivalent tournament in 2006, Division Three of the African region of the World Cricket League, it finished as runner-up to Mozambique, and just missed a promotion to Division Two.

Basketball

The Sierra Leone national basketball team represents Sierra Leone in international men's basketball competitions and is controlled by the Sierra Leone Basketball Federation. The squad is mostly home-based, with a few foreign players.

14

South-West Africa

West Africa or Western Africa is the western most region of the African continent. Geopolitically, the UN definition of Western Africa includes the following 15 countries & an area of approximately 5 million square km:

- Benin
- Burkina Faso
- Ivory Coast
- Cape Verde
- The Gambia
- Ghana
- Guinea
- Guinea-Bissau
- Liberia
- Mali
- Mauritania
- Niger
- Nigeria
- Senegal
- Sierra Leone
- Togo.

With the exception of Mauritania, all of these countries are members of the ECOWAS or Economic Community of West African States. The UN region also includes the island of Saint Helena, a British overseas territory in the South Atlantic Ocean.

Background

West Africa is west of an imagined north-south axis lying close to 10° east longitude. The Atlantic Ocean forms the western and southern borders of the region. The northern border is the Sahara Desert, with the Ranishanu Bend generally considered the northernmost part of the region. The eastern border is less precise, with some placing it at the Benue Trough, and others on a line running from Mount Cameroon to Lake Chad.

Colonial boundaries are reflected in the modern boundaries between contemporary West African nations, cutting across ethnic and cultural lines, often dividing single ethnic groups between two or more countries.

The inhabitants of West Africa are, in contrast to most of Southern and Middle Africa, non-Bantu speaking peoples.

Geography and Climate

West Africa, if one includes the western portion of the Maghreb (Western Sahara, Morocco, Algeria, and Tunisia), occupies an area in excess of 6,140,000 km, or approximately one-fifth of Africa. The vast majority of this land is plains lying less than 300 meters above sea level, though isolated high points exist in numerous countries along the southern shore of the region.

The northern section of West Africa is composed of semi-arid terrain known as Sahel, a transitional zone between the Sahara and the savannahs of the western Sudan forests form a third belt between the savannas and the southern coast, ranging from 160 km to 240 km in width.

Culture

Despite the wide variety of cultures in West Africa, from Nigeria through to Senegal, there are general similarities in dress, cuisine, music and culture that are not shared extensively with groups outside the geographic region.

Religion

Islam is the predominant historical religion of the West African interior and the far west coast of the continent; Christianity is the predominant religion in coastal regions of

Nigeria, Ghana, and Cote d'Ivoire; and elements of indigenous religions are practiced throughout.

Along with historic migrations, these religions have culturally linked the peoples of West Africa more than those in other parts of Sub-Saharan Africa. Traditionally, musical and oral history as conveyed over generations by Griots are typical of West African culture.

Recreation

The game Oware is quite popular in many parts of West Africa. Soccer is also a pastime enjoyed by many, either spectating or playing. The national teams of some West African nations, especially Nigeria, Ghana and the Ivory Coast, regularly qualify for the World Cup.

Music

Mbalax, Highlife, Fuji and Afrobeat are all modern musical genres which enjoin listeners in this region.

A typical formal attire worn in this region is the flowing Boubou (also known as *Agbada* and *Babariga*), which has its origins in the clothing of nobility of various West African empires in the 12th century.

The Djembe drum, whose origins lie with the Mandinka peoples, is now a popularly played drum among many West African ethnic groups. The Kora is a 21-string harp-lute of Mandinkan origin, played by various groups in the region. The Djembe, Kora, the silk Kente cloth of the Akan peoples of Ghana and the distinct Sudano-Sahelian architectural style seen in the many mosques of the region, are the primary symbolic icons of West African culture.

History

The history of West Africa can be divided into five major periods: first, its prehistory, in which the first human settlers arrived, developed agriculture, and made contact with peoples to the north; the second, the Iron Age empires that consolidated both intra-African, and extra-African trade, and developed centralized states; third, Major polities flourished, which would undergo an extensive history of contact with non-Africans; fourth, the brief colonial period, in which Great Britain and

France controlled nearly the whole of the region; fifth, the post-independence era, in which the current nations were formed.

Prehistory

Early human settlers arrived in West Africa around 12,000 B.C. Sedentary farming began in, or around the fifth millennium B.C, as well as the domestication of cattle. By 400 B.C, iron working technology allowed an expansion of agricultural productivity, and the first city-states formed.

The domestication of the camel allowed the development of a cross-Saharan trade with cultures across the Sahara, including Carthage and the Berbers; major exports included gold, cotton cloth, metal ornaments and leather goods, which were then exchanged for salt, horses, textiles, and other such materials. Local leather, cloth, and gold also contributed to the abundancy of prosperity for many of the following empires.

Empires

The development of the region's economy allowed more centralized states and civilizations to form, beginning with the Nok civilization which began 500 B.C. and the Ghana Empire in the 8th century AD which stretched to the Mali empire. Based on the city of Kumbi Saleh in modern-day Mauritania, the empire came to dominate much of the region until its defeat by Almoravid invaders in 1052.

The Sosso Empire sought to fill the void, but was defeated (c. 1240) by the Mandinka forces of Sundiata Keita, founder of the new Mali Empire. The Mali Empire continued to flourish for several centuries, most particularly under Sundiata's grandnephew) Musa I, before a succession of weak rulers led to its collapse under Mossi, Tuareg and Songhai invaders. In the fifteenth century, the Songhai would form a new dominant state based on Gao, in the Songhai Empire, under the leadership of Sonni Ali and Askia Mohammed.

Further south, Osei Tutu and Okomfo Anokye have started to build the Empire of Ashanti Meanwhile, south of the Sudan, strong city states arose in Igboland, such as the Kingdom of Nri in the tenth century, while Ife, Bono, and Benin arose around the fourteenth century. Further east, Oyo arose as the dominant Yoruba state and the Aro Confederacy as a dominant Igbo state in modern-day Nigeria.

Slavery and European Contact

Following the 1591 destruction of the Songhai capital by Moroccan invaders, a number of smaller states arose across West Africa, including the Bambara Empire of Segou, the Bambara kingdom of Kaarta, the Peul/Malinke kingdom of Khasso, and the Kenedougou Empire of Sikasso.

Portuguese traders began establishing settlements along the coast in 1445, followed by the French and English; the African slave trade began not long after, which over the following centuries would debilitate the region's economy and population. The slave trade also encouraged the formation of states such as the Asante Empire, Bambara Empire and Dahomey, whose economies largely depended on exchanging slaves for European firearms, which were then used to capture more slaves.

The expanding trans-Atlantic slave trade produced significant populations of West Africans living in the New World, recently colonized by Europeans. The oldest known remains of African slaves in the Americas were found in Mexico in early 2006; they are thought to date from the late 16th century and the mid-17th century.

European and American governments passed legislation prohibiting the Atlantic slave trade in the 19th century, though slavery in the Americas persisted in some capacity through the century; the last country to abolish the institution was Brazil in 1888. Descendants of West Africans make up large and important segments of the population in Brazil, the Caribbean, Latin America, and the United States.

Colonialism

In the early nineteenth century, a series of Fulani reformist jihads swept across Western Africa. The most notable include Usman dan Fodio's Fulani Empire, which replaced the Hausa city-states, Seku Amadu's Massina Empire, which defeated the Bambara, and El Hadj Umar Tall's Toucouleur Empire, which briefly conquered much of modern-day Mali. However, the French and British continued to advance in the Scramble for

Africa, subjugating kingdom after kingdom. With the fall of Samory Ture's new-founded Wassoulou Empire in 1898 and the Ashanti queen Yaa Asantewaa in 1902, most West African military resistance to colonial rule came to an effective end.

Britain controlled The Gambia, Sierra Leone, Ghana, and Nigeria throughout the colonial era, while France unified Senegal, Guinea, Mali, Burkina Faso, Benin, Cote d'Ivoire and Niger into French West Africa.

Portugal founded the colony of Guinea-Bissau, while Germany claimed Togoland, but was forced to divide it between France and Britain following First World War in order of the treaty of versallies. Only Liberia retained its independence, at the price of major territorial concessions.

Postcolonial Era

Following World War II, nationalist movements arose across West Africa. In 1957, Ghana, under Kwame Nkrumah, became the first sub-Saharan colony to achieve its independence, followed the next year by France's colonies (Guinea in 1958 under the leadership of President Ahmed Sekou Toure); by 1974, West Africa's nations were entirely autonomous.

Since independence, many West African nations have been plagued by corruption and instability, with notable civil wars in Nigeria, Sierra Leone, Liberia, and Cote d'Ivoire, and a succession of military coups in Ghana and Burkina Faso. Many states have failed to develop their economies despite enviable natural resources, and political instability is often accompanied by undemocratic government.

AIDS is also a growing problem for the region, particularly in Cote d'Ivoire and Liberia. Famine has been a problem in parts of northern Mali and Niger, the latter of which is currently undergoing a food crisis.

Until recently, most governments in West Africa were illiberal and corrupt and several countries have been plagued with political coups, ethnic violence and oppressive dictators. Since the end of colonialism, the region has been the stage for some of the most brutal conflicts ever to erupt. Among the latter are:

- Nigerian Civil War

- First Liberian Civil War
- Second Liberian Civil War
- Guinea-Bissau Civil War
- Ivorian Civil War
- Sierra Leone Civil War.

Though a few countries like Ghana and Senegal have enjoyed relative stability and growth, all progress in the region is contingent on the efficacy and justness of governance and the fair allocation of resources which, for the moment, both leave much to be desired.

Regional Organizations

The Economic Community of West African States (ECOWAS), founded by the 1975 Treaty of Lagos, is an organization of West African states which aims to promote the region's economy. The West African Monetary Union (or UEMOA from its name in French, *Union economique et monetaire ouest-africaine*) is limited to the eight, mostly Francophone countries that employ the CFA franc as their common currency. The Liptako-Gourma Authority of Mali, Niger, and Burkina Faso seeks to jointly develop the contiguous areas of the three countries.

15

South Africa

The Republic of South Africa is a country located at the southern tip of Africa, with a 2,798 kilometres (1,739 mi) coastline on the Atlantic and Indian Oceans. To the north lie Namibia, Botswana and Zimbabwe; to the east are Mozambique and Swaziland; while Lesotho is an independent country wholly surrounded by South African territory.

South Africa is known for a diversity in cultures and languages. Eleven official languages are recognised in the constitution. Two of these languages are of European origin: Afrikaans, a language which originated mainly from Dutch that is spoken by the majority of white and Coloured South Africans, and South African English. Though English has a large role in public and commercial life, it is nevertheless only the fifth most-spoken home language.

South Africa is ethnically diverse. About 79.5% of the South African population is of black African ancestry, divided among a variety of ethnic groups speaking different Bantu languages, nine of which have official status. South Africa also contains the largest European, Indian, and racially mixed communities in Africa. About a quarter of the population is unemployed and lives on less than US $1.25 a day.

South Africa is a constitutional democracy in the form of a parliamentary republic with a parliamentarily-dependent head of state. It is one of the founding members of the African Union, and has the largest economy of all the members. It is also a founding member of the United Nations and NEPAD.

South Africa is a member of the Commonwealth of Nations, Antarctic Treaty System, Group of 77, South Atlantic Peace and Cooperation Zone, Southern African Customs Union, World Trade Organization, International Monetary Fund, G20 and G8+5.

History

South Africa contains some of the oldest archaeological sites in the world. Extensive fossil remains at the Sterkfontein, Kromdraai and Makapansgat caves suggest that various australopithecines existed in South Africa from about three million years ago. These were succeeded by various species of *Homo*, including *Homo habilis*, *Homo erectus* and modern humans, *Homo sapiens*.

Settlements of Bantu-speaking peoples, who were iron-using agriculturists and herdsmen, were already present south of the Limpopo River by the fourth or fifth century CE. They displaced, conquered and absorbed the original Khoisan speakers. The Bantu slowly moved south. The earliest ironworks in modern-day Kwa Zulu-Natal Province are believed to date from around 1050. The southern most group was the Xhosa people, whose language incorporates certain linguistic traits from the earlier Khoisan people. The Xhosa reached the Great Fish River, in today's Eastern Cape Province. As they migrated, these larger Iron Age populations displaced or assimilated earlier peoples, who often had hunter-gatherer societies.

Modern humans have inhabited Southern Africa for more than 100,000 years. At the time of European contact, the dominant indigenous peoples were tribes who had migrated from other parts of Africa about one thousand years before. From the 4th-5th century CE, Bantu-speaking tribes had steadily moved south, where they displaced, conquered and assimilated original Khoikhoi and San peoples of southern Africa. At the time of European contact, the two major groups were the Xhosa and Zulu peoples.

Great Britain took over the Cape of Good Hope area in 1795, ostensibly to stop it from falling under control of the French First Republic, which had invaded the Dutch Republic. Given its standing interests in Australia and India, Great

Britain wanted to use Cape Town as an interim port for its merchants' long voyages. The British returned Cape Town to the Dutch in 1803, but soon afterwards the Dutch East India Company declared bankruptcy.

The British annexed the Cape Colony in 1806. The British continued the frontier wars against the Xhosa, pushing the eastern frontier eastward through a line of forts established along the Fish River. They consolidated the territory by encouraging British settlement. Due to pressure of abolitionist societies in Britain, the British parliament first stopped its global slave trade with the passage of the Slave Trade Act 1807, then abolished slavery in all its colonies with the Slavery Abolition Act 1833.

In the first two decades of the 19th century, the Zulu people grew in power and expanded their territory under their leader, Shaka. Shaka's depredations led indirectly to the Mfecane ("Crushing") that devastated the inland plateau in the early 1820s. An offshoot of the Zulu, the Matabele, created an even larger empire under their king Mzilikazi, including large parts of the highveld.

During the 1830s, approximately 12,000 Boers (later known as Voortrekkers), departed from the Cape Colony, where they had been subjected to British control. They migrated to the future Natal, Orange Free State and Transvaal regions. The Boers founded the Boer Republics: the South African Republic (now Gauteng, Limpopo, Mpumalanga and North West provinces) and the Orange Free State.

The discovery of diamonds in 1867 and gold in 1884 in the interior encouraged economic growth and immigration. This intensified the European-South African subjugation of the indigenous people. The struggle to control these important economic resources was a factor between Europeans and the indigenous population, and also between the Boers and the British.

The Boer Republics successfully resisted British encroachments during the First Boer War (1880–1881) using guerrilla warfare tactics, which were well suited to local conditions. However, the British returned with greater numbers, more experience, and more suitable tactics in the Second Boer War (1899–1902), which was won by the British.

20th Century

Within the country, anti-British policies among white South Africans focused on independence. During the Dutch and British colonial years, racial segregation was mostly informal, though some legislation were enacted to control the settlement and movement of native people, including the Native Location Act of 1879 and the system of *pass laws*. Power was held by the European colonists.

After four years of negotiating, the Union of South Africa was created from the Cape and Natal colonies, as well as the republics of Orange Free State and Transvaal, on 31 May 1910, exactly eight years after the end of the Second Boer War. The newly created Union of South Africa was a dominion of Great Britain. The Natives' Land Act of 1913 severely restricted the ownership of land by 'blacks'; at that stage they had control of a mere 7% of the country. The amount of land reserved for indigenous peoples was later marginally increased.

In the Boer republics, from as early as the Pretoria Convention and subsequent South African governments, the system became legally institutionalised segregation, later known as *apartheid*. The government established three classes of racial stratification: white, coloured, and black, with rights and restrictions for each. In 1931 the union was effectively granted independence from the United Kingdom with the passage of the Statute of Westminster. In 1934, the South African Party and National Party merged to form the United Party, seeking reconciliation between Afrikaners and English-speaking "Whites". In 1939 the party split over the entry of the Union into World War II as an ally of the United Kingdom, a move which the National Party followers strongly opposed.

In 1948, the National Party was elected to power. It intensified the implementation of racial segregation begun under Dutch and British colonial rule, and subsequent South African governments since the Union was formed. The Nationalist Government systematised existing segregationist laws, classifying all peoples into three races, developing rights and limitations for each, such as pass laws and residential restrictions. The white minority controlled the vastly larger black majority. The system of segregation became known collectively as *apartheid.*

While the White minority enjoyed the highest standard of living in all of Africa, comparable to First World western nations, the Black majority remained disadvantaged by almost every standard, including income, education, housing, and life expectancy. On 31 May 1961, following a whites-only referendum, the country became a republic and left the Commonwealth. Queen Elizabeth II ceased to be head of state, and the last Governor-General became State President.

Despite opposition both within and outside the country, the government legislated for a continuation of apartheid. As the 20th century went on, Apartheid became increasingly controversial, some Western nations and institutions began to boycott doing business with the country because of its racial policies and oppression of civil rights leading to widespread international sanctions, divestment and growing unrest and oppression within South Africa. A long period of harsh suppression by the government, and at times violent resistance, strikes, marches, protests, and sabotage by bombing and other means, by various anti-apartheid movements, most notably the African National Congress (ANC), followed.

In the late 1970s, South Africa began a programme of nuclear weapons development. In the following decade, it produced six deliverable nuclear weapons.

The Mahlabatini Declaration of Faith, signed by Mangosuthu Buthelezi and Harry Schwarz in 1974, enshrined the principles of peaceful transition of power and equality for all, the first of such agreements by acknowledged black and white political leaders in South Africa, which would ultimately end with the negotiations between F.W. de Klerk and Nelson Mandela in 1993.

After years of internal protests, activism and insurgency by black South Africans and their allies, finally in 1990 the National Party government took the first step towards dismantling discrimination when it lifted the ban on the African National Congress and other political organisations. It released

Nelson Mandela from prison after twenty-seven years' incarceration on a sabotage sentence. A negotiation process followed. The government repealed apartheid legislation. South Africa destroyed its nuclear arsenal and acceded to the Nuclear Non-Proliferation Treaty. South Africa held its first universal elections in 1994, which the ANC won by an overwhelming majority. It has been in power ever since. The country then rejoined the Commonwealth of Nations.

In post-apartheid South Africa, unemployment has been extremely high. While many blacks have risen to middle or upper classes, the overall unemployment rate of blacks worsened between 1994 and 2003. Poverty among whites, previously rare, increased. While some have attributed this partly to the legacy of the apartheid system, increasingly many attribute it to the failure of the current government's policies. In addition, the current government has struggled to achieve the monetary and fiscal discipline to ensure both redistribution of wealth and economic growth. Since the ANC-led government took power, the United Nations Human Development Index of South Africa has fallen, while it was steadily rising until the mid-1990s. Some of this could possibly be attributed to the AIDS pandemic and the failure of the government to take steps to address it.

Government and Politics

South Africa has three capital cities: Cape Town, the largest of the three, is the *legislative* capital; Pretoria is the *administrative* capital; and Bloemfontein is the *judicial* capital. South Africa has a bicameral parliament: the National Council of Provinces (the upper house) has 90 members, while the National Assembly (the lower house) has 400 members.

Members of the lower house are elected on a population basis by proportional representation: half of the members are elected from national lists and the other half are elected from provincial lists. Ten members are elected to represent each province in the National Council of Provinces, regardless of the population of the province. Elections for both chambers are held every five years. The government is formed in the lower house, and the leader of the majority party in the National Assembly is the President.

The primary sources of South Africa law are Roman-Dutch mercantile law and personal law with English Common law, as imports of Dutch settlements and British colonialism. The first European based law in South Africa was brought by the Dutch East India Company and is called Roman-Dutch law. It was imported before the codification of European law into the Napoleonic Code and is comparable in many ways to Scots law. This was followed in the 19th century by English law, both common and statutory. Starting in 1910 with unification, South Africa had its own parliament which passed laws specific for South Africa, building on those previously passed for the individual member colonies. During the years of apartheid, the country's political scene was dominated by figures like B. J. Vorster and P. W. Botha, as well as opposition figures such as Harry Schwarz, Joe Slovo and Helen Suzman.

Since the end of apartheid in 1994, South African politics have been dominated by the African National Congress (ANC), which has been the dominant party with 60–70% of the vote. The main challenger to the rule of the ANC is the Democratic Alliance party, which received 16.7% of the vote in the 2009 election and 14.8% in the 2006 election.

The formerly dominant New National Party, which introduced apartheid through its predecessor, the National Party, chose to merge with the ANC on 9 April 2005. Other major political parties represented in Parliament are the Congress of the People, which split from the ANC and won 7.4% of the vote in 2009, and the Inkatha Freedom Party, which mainly represents Zulu voters and took 4.6% of the vote in the 2009 election.

Since 2004, the country has had many thousands of popular protests, some violent, making it, according to one academic, the "most protest-rich country in the world". Many of these protests have been organised from the growing shanty towns that surround South African cities.

In 2008, South Africa placed 5th out of 48 sub-Saharan African countries on the Ibrahim Index of African Governance. South Africa scored well in the categories of Rule of Law, Transparency & Corruption and Participation & Human Rights, but was let down by its relatively poor performance in Safety

& Security. The Ibrahim Index is a comprehensive measure of African governance, based on a number of different variables which reflect the success with which governments deliver essential political goods to its citizens.

After the end of apartheid in 1994, the "independent" and "semi-independent" Bantustans were integrated into the political structure of South Africa by the abolition of the four former provinces (Cape Province, Natal Province, Orange Free State and Transvaal Province) and the creation of nine fully integrated new provinces. The generally smaller size of the new provinces theoretically means that local governments have more resources to distribute over smaller areas. The provinces are subdivided into 52 districts: 6 metropolitan and 46 district municipalities. The district municipalities are further subdivided into 231 local municipalities. The metropolitan municipalities perform the functions of both district and local municipalities.

Foreign Relations and Military

Since the end of apartheid, the South African foreign policy has focused on its African partners particularly in the Southern African Development Community (SADC) and the African Union. South Africa has played a key role as a mediator in African conflicts over the last decade, such as in Burundi, the Democratic Republic of Congo, the Comoros, and Zimbabwe. After apartheid ended, South Africa was readmitted to the Commonwealth of Nations.

As the Union of South Africa, South Africa was a founding member of the United Nations. The then Prime Minister Jan Smuts wrote the preamble to the United Nations Charter. South Africa was a non-permanent member of the United Nations Security Council between 2007 and 2008, and has attracted controversy by voting against a resolution criticising the Burmese government in 2006 and against the implementation of sanctions against Zimbabwe in 2008. South Africa is a member of the Group of 77 and chaired the organisation in 2006. South Africa is a member of the South Atlantic Peace and Cooperation Zone, Southern African Customs Union, World Trade Organization, International Monetary

Fund, G20 and G8+5. The South African National Defence Force (SANDF) was created in 1994, as an all volunteer force composed of as the former South African Defence Force, the forces of the African nationalist groups (Umkhonto we Sizwe and Azanian People's Liberation Army), and the former Bantustan defence forces. The SANDF is subdivided into four branches, the South African Army, the South African Air Force, the South African Navy, and the South African Medical Service.

In recent years, the SANDF has become a major peacekeeping force in Africa, and has been involved in operations in Lesotho, the Democratic Republic of the Congo, and Burundi, amongst others. It has also participated as a part of multinational UN peacekeeping forces.

South Africa undertook a nuclear weapons programme in the 1970s and may have conducted a nuclear test over the Atlantic in 1979. It is the only African country to have successfully developed nuclear weapons. It has become the first country (followed by Ukraine) with nuclear capability to voluntarily renounce and dismantle its programme and in the process signed the Nuclear Non-Proliferation Treaty in 1991.

Geography

South Africa is located at the southern most region of Africa, with a long coastline that stretches more than 2,500 km (1,553 mi) and along two oceans (the South Atlantic and the Indian). At 1,219,912 km (471,011 sq mi), South Africa is the 25th-largest country in the world and is comparable in size to Colombia. Njesuthi in the Drakensberg at 3,408 m (11,181 ft) is the highest peak in South Africa.

The interior of South Africa is a vast, flat, and sparsely populated scrubland, the Karoo, which is drier towards the northwest along the Namib desert. In contrast, the eastern coastline is lush and well-watered, which produces a climate similar to the tropics.

To the north of Johannesburg, the altitude drops beyond the escarpment of the Highveld, and turns into the lower lying Bushveld, an area of mixed dry forest and an abundance of wildlife. East of the Highveld, beyond the eastern escarpment, the Lowveld stretches towards the Indian Ocean. It has particularly high temperatures, and is also the location of extended subtropical agriculture.

South Africa also has one possession, the small sub-Antarctic archipelago of the Prince Edward Islands, consisting of Marion Island (290 km/110 sq mi) and Prince Edward Island (45 km/17 sq mi) (not to be confused with the Canadian province of the same name).

Climate

South Africa has a generally temperate climate, due in part to being surrounded by the Atlantic and Indian Oceans on three sides, by its location in the climatically milder southern hemisphere and due to the average elevation rising steadily towards the north (towards the equator) and further inland. Due to this varied topography and oceanic influence, a great variety of climatic zones exist.

The climatic zones vary, from the extreme desert of the southern Namib in the farthest northwest to the lush subtropical climate in the east along the Mozambique border and the Indian ocean. From the east, the land quickly rises over a mountainous escarpment towards the interior plateau known as the Highveld. Even though South Africa is classified as semi-arid, there is considerable variation in climate as well as topography.

The extreme southwest has a climate remarkably similar to that of the Mediterranean with wet winters and hot, dry summers, hosting the famous Fynbos Biome of grassland and thicket. This area also produces much of the wine in South Africa. This region is also particularly known for its wind, which blows intermittently almost all year. The severity of this wind made passing around the Cape of Good Hope particularly treacherous for sailors, causing many shipwrecks. Further east on the south coast, rainfall is distributed more evenly throughout the year, producing a green landscape. This area is popularly known as the Garden Route.

The Free State is particularly flat due to the fact that it lies centrally on the high plateau. North of the Vaal River, the Highveld becomes better watered and does not experience subtropical extremes of heat. Johannesburg, in the centre of the Highveld, is at 1,740 m (5,709 ft) and receives an annual rainfall of 760 mm (29.9 in). Winters in this region are cold, although snow is rare.

The high Drakensberg mountains, which form the south-eastern escarpment of the Highveld, offer limited skiing opportunities in winter. The coldest place in South Africa is Sutherland in the western Roggeveld Mountains, where midwinter temperatures can reach as low as "15 °C (5.0 °F). The deep interior has the hottest temperatures: a temperature of 51.7 °C (125.06 °F) was recorded in 1948 in the Northern Cape Kalahari near Upington.

Flora and Fauna

South Africa is ranked sixth out of the world's seventeen megadiverse countries, with more than 20,000 different plants, or about 10% of all the known species of plants on Earth, making it particularly rich in plant biodiversity. The most prevalent biome in South Africa is the grassland, particularly on the Highveld, where the plant cover is dominated by different grasses, low shrubs, and acacia trees, mainly camel-thorn and whitethorn. Vegetation becomes even more sparse towards the northwest due to low rainfall. There are several species of water-storing succulents like aloes and euphorbias in the very hot and dry Namaqualand area. The grass and thorn savannah turns slowly into a bush savannah towards the north-east of the country, with denser growth. There are significant numbers of baobab trees in this area, near the northern end of Kruger National Park.

The Fynbos Biome, which makes up the majority of the area and plant life in the Cape floristic region, one of the six floral kingdoms, is located in a small region of the Western Cape and contains more than 9,000 of those species, making it among the richest regions on earth in terms of floral biodiversity. The majority of the plants are evergreen hard-leaf plants with fine, needle-like leaves, such as the sclerophyllous plants. Another uniquely South African plant is the protea genus of flowering plants. There are around 130 different species of protea in South Africa. While South Africa has a great wealth of flowering plants, only 1% of South Africa is forest, almost exclusively in the humid coastal plain of Kwa Zulu-Natal, where there are also areas of Southern Africa mangroves in river mouths. There are even smaller reserves of forests that are out of the reach of fire, known as montane forests. Plantations of imported tree species are predominant, particularly the non-native eucalyptus and pine. South Africa has lost a large area of natural habitat in the last four decades, primarily due to overpopulation, sprawling development patterns and deforestation during the nineteenth century. South Africa is one of the worst affected countries in the world when it comes to invasion by alien species with many (e.g. Black Wattle, Port Jackson, Hakea, Lantana and Jacaranda) posing a significant threat to the native biodiversity and the already scarce water resources. The original temperate forest found by the first European settlers was exploited ruthlessly until only small patches remained. Currently, South African hardwood trees like Real Yellowwood *(Podocarpus latifolius)*, stinkwood *(Ocotea bullata)*, and South African Black Ironwood *(Olea laurifolia)* are under government protection.

Numerous mammals are found in the bushveld including lions, leopards, white rhinos, blue wildebeest, kudus, impalas, hyenas, hippopotamus and giraffes. A significant extent of the bushveld exists in the north-east including Kruger National Park and the Mala Mala Reserve, as well as in the far north in the Waterberg Biosphere.

Climate change is expected to bring considerable warming and drying to much of this already semi-arid region, with greater frequency and intensity of extreme weather events such as heat waves, flooding and drought. According to computer generated climate modelling produced by the South African National Biodiversity Institute parts of southern Africa will see an increase in temperature by about one degree Celsius along the coast to more than four degrees Celsius in the already hot hinterland such as the Northern Cape in late spring and summertime by 2050.

The Cape Floral Kingdom has been identified as one of the global biodiversity hotspots since it will be hit very hard by climate change and has such a great diversity of life. Drought, increased intensity and frequency of fire and climbing temperatures are expected to push many of these rare species towards extinction.

South Africa houses many endemic species, among them the critically endangered Riverine Rabbit (*Bunolagus monticullaris*) in the Karoo.

Economy

The economy of South Africa has a two tiered economy; one rivalling other developed countries and the other with only the most basic infrastructure. It is therefore a productive and industrialised economy that exhibits many characteristics associated with developing countries, including a division of labour between formal and informal sectors and an uneven distribution of wealth and income. The primary sector, based on manufacturing, services, mining, and agriculture, is well developed.

South Africa's transportation infrastructure is among the best in Africa, supporting both domestic and regional needs. OR Tambo International Airport serves as a hub for flights to other Southern African and International countries. South Africa also has several major ports that make it the central point for most trade in the Southern African region.

History

This is a chart of the trend of South Africa's gross domestic product at market prices estimated by the International Monetary Fund with figures in millions of South African rand.

Year	*Gross Domestic Product*	*US Dollar Exchange*	*Per Capita Income (as % of USA)*
1980	62,730	0.77 Rand	22.55
1985	127,598	1.47 Rand	9.80
1990	289,816	2.58 Rand	13.09
1995	1,548,100	3.62 Rand	13.27
2000	922,148	6.93 Rand	8.58
2005	1,523,254	6.36 Rand	12.32

The formal economy of South Africa has its beginnings in the arrival of Dutch settlers in 1652, originally sent by the Dutch East India Company to establish a provisioning station for passing ships. As the colony increased in size, with the arrival of French Huguenots and German citizens, some of the colonists were set free to pursue commercial farming, leading to the dominance of agriculture in the economy.

At the end of the 18th century, the British gained control of the colony, imposing the English language on the colonists, who were now developing a culture of their own. This in turn lead to the Great Trek, spreading farming deeper into the mainland, as well as the establishment of the independent Boer Republics of Transvaal and the Orange Free State.

In 1870 diamonds were discovered in Kimberley, while in 1886 some of the world's largest gold deposits were discovered in the Witwatersrand region of Transvaal, quickly transforming the economy into a resource-dominated one. The British, seeking the riches of the gold fields, invaded the Boer republics and gained control of them in 1902 after the Second Boer War. The country also entered a period of industrialisation during this time, including the organisation of the first South African trade unions. The government soon started putting laws distinguishing between different races in place. In 1948 the National Party won the national elections, and immediately started implementing an even stricter race-based policy named Apartheid, effectively dividing the economy into a privileged white one, and an impoverished black one. The policy was widely criticized and led to crippling sanctions being placed against the country in the 1980s.

South Africa held its first multi-racial elections in 1994, leaving the newly-elected African National Congress (ANC) government the daunting task of trying to restore order to an economy harmed by sanctions, while also integrating the previously-disadvantaged segment of the population into it. As of 2005, agriculture contributes only 3.4% to the country's GDP, while services now account for 65.1%.

In April 2009, amidst fears that South Africa would soon join much of the rest of the world in recession, Reserve Bank Governor Tito Mboweni and Finance Minister differed on the matter: whereas Manuel foresaw a quarter of economic growth, Mboweni predicted further decline: "technically," he said, "that's a recession."

Sectors

Telecommunications

The domestic telecommunications infrastructure provides modern and efficient service to urban areas, including cellular and internet services. In 1997, Telkom, the South African telecommunications parastatal, was partly privatised and entered into a strategic equity partnership with a consortium of two companies, including SBC, a U.S. telecommunications company. In exchange for exclusivity (a monopoly) to provide certain services for 5 years, Telkom assumed an obligation to facilitate network modernisation and expansion into unserved areas. A Second Network Operator was to be licensed to compete with Telkom across its spectrum of services in 2002, although this license was only officially handed over in late 2005 and has recently begun operating under the name, Neotel. Four cellular companies provide service to over 20 million subscribers, with South Africa considered to have the 4th most advanced mobile telecommunications network worldwide. The four cellular providers are Vodacom, MTN, Cell C and Virgin Mobile.

Agriculture

Unlike other African countries, South Africa's agricultural sector is not dominated by subsistence farming, with most farms being large commercial, albeit family-owned, enterprises. The country is completely self-reliant and has more than enough output to export massive amounts of agricultural produce. Many other southern African countries rely on South Africa for maize imports.

Due to the country's varied climate, many different crops are grown. The Western Cape province has the most varied and prolific agricultural sector. Wine has become a massive export, with South Africa now being the 5th largest producer worldwide. Deciduous fruit is also of major importance, with grapes, apples, cherries, pears, peaches, citrus and other fruit being exported in great quantities, mostly to Europe. Heavy wheat cultivation also occurs in the region, along with major wheat growing areas in the Highveld of Mpumalanga and the Free State. The Free State is the leading producer of South Africa's staple, maize.

The vast inland regions of the Karoo provide ideal conditions for livestock farming, especially sheep farming (for wool and mutton). Cattle farming is more popular amongst the indigenous people and flourishes more in the more well-watered eastern areas of South Africa. Ostrich farming is popular in the Oudtshoorn area of the Western Cape, along with extensive dairy farming in the Garden Route area just to the south. Sugarcane farming is a mainstay on the Kwa Zulu-Natal coast, with subtropical fruits, such as mangos, lychees, papaya, bananas and melons being extensively cultivated in Kwa Zulu-Natal, Limpopo and Mpumalanga Lowveld areas. Pineapples are cultivated around East London. Many game farms specializing in South African wild antelope are also gaining in importance and are found mainly in the north and east of South Africa.

Despite attempts by government to reform the distribution of land, historically mostly held by whites, these efforts have not yet translated into growth in the agricultural sector, which continues to lag or decline in relation to the rest of the economy. This may also be due to the fact that indigenous people are mostly subsistence farmers and that anti-competitive practices like agricultural subsidies in developed countries and climate change are curtailing sector growth.

According to the OECD, "Agriculture contributes less than 4% to GDP but accounts for 10% of total reported employment."

South Africa has a large agricultural sector and is a net exporter of farming products. There are almost a thousand agricultural cooperatives and agribusinesses throughout the country, and agricultural exports have constituted 8% of South African total exports for the past five years. The agricultural industry contributes around 10% of formal employment, relatively low compared to other parts of Africa, as well as providing work for casual labourers and contributing around 2.6% of GDP for the nation. However, due to the aridity of the land, only 13.5% can be used for crop production, and only 3% is considered high potential land.

Although the commercial farming sector is relatively well developed, people in some rural areas still survive on subsistence agriculture. It is the eighth largest wine producer in the world, and the eleventh largest producer of sunflower seed. South Africa is a net exporter of agricultural products and foodstuffs, the largest number of exported items being sugar, grapes, citrus, nectarines, wine and deciduous fruit. The largest locally produced crop is maize (corn), and it has been estimated that 9 million tons are produced every year, with 7.4 million tons being consumed. Livestock are also popular on South African farms, with the country producing 85% of all meat consumed. The dairy industry consists of around 4,300 milk producers providing employment for 60,000 farm workers and contributing to the livelihoods of around 40,000 others.

In recent years, the government has introduced several agricultural sector reforms, such as land reform and the deregulation of the market for agricultural products. The South African government has set a target of transferring 30% of productive farmland from whites to 'previously disadvantaged' blacks by 2014. Land reform has been criticised both by farmers' groups and by landless workers, the latter alleging that the pace of change has not been fast enough, and the former alleging 'racist' treatment and expressing concerns that a similar situation to Zimbabwe's land reform policy may develop, a fear exacerbated by comments made by former deputy president Phumzile Mlambo-Ngcuka. The sector continues to face problems, with increased foreign competition and crime being two of the major challenges for the industry. The government has been accused of either putting in too much effort, or not enough effort, to tackle the problem of farm attacks as opposed to other forms of violent crime.

Another issue which affects South African agriculture is environmental damage caused by misuse of the land and global climate change. South Africa is unusually vulnerable to climate change and resultant diminution of surface waters. Some predictions show surface water supply could decrease by 60% by the year 2070 in parts of the Western Cape. To reverse the damage caused by land mismanagement, the government has supported a scheme which promotes sustainable development and the use of natural resources. Maize production, which contributes to a 36% majority of the gross value of South Africa's field crops, has also experienced negative effects due to climate change. The estimated value of loss, which takes into consideration scenarios with and without the carbon dioxide fertilisation effect, ranges between 10s and 100s of millions of Rands.

Trade and Investment

South Africa has rich mineral resources. It is the world's largest producer and exporter of platinum and, until 2007, gold, and also exports a significant amount of coal. Another major export is diamonds. During 2000, platinum overtook gold as South Africa's largest foreign exchange earner. The value-added processing of minerals to produce ferroalloys, stainless steels, and similar products is a major industry and an important growth area. The country's diverse manufacturing industry is a world leader in several specialised sectors, including railway rolling stock, synthetic fuels, and mining equipment and machinery.

Agriculture, based on a 2005 estimate by The World Factbook, accounts for only 3.4% of the gross domestic product. Major crops include citrus and deciduous fruits, corn, wheat, dairy products, sugarcane, tobacco, wine and wool. South Africa has many developed irrigation schemes and is a net exporter of food.

Exports reached 29.1% of GDP in 2001, up from 11.5% a decade ago. South Africa's major trading partners include the United Kingdom, the United States, Germany, Italy, Belgium, China, and Japan. South Africa's trade with other Sub-Saharan African countries, particularly those in the Southern Africa region, has increased substantially. South Africa is a member of the Southern African Customs Union (SACU) and the Southern African Development Community (SADC). In August 1996, South Africa signed a regional trade protocol agreement with its SADC partners. The agreement was ratified in December 1999 and implementation began in September 2000. It intends to provide duty-free treatment for 85% of trade by 2008 and 100% by 2012.

South Africa has made great progress in dismantling its old economic system, which was based on import substitution, high tariffs and subsidies, anticompetitive behaviour, and extensive government intervention in the economy. The new leadership has moved to reduce the government's role in the economy and to promote private sector investment and competition. It has significantly reduced tariffs and export subsidies, loosened exchange controls, cut the secondary tax on corporate dividends, and improved enforcement of intellectual property laws. A new competition law was passed and became effective on 1 September 1999. A U.S.-South Africa bilateral tax treaty went into effect on 1 January 1998, and a bilateral trade and investment framework agreement was signed in February 1999.

South Africa is a member of the World Trade Organization (WTO). U.S. products qualify for South Africa's most-favoured-nation tariff rates. South Africa also is an eligible country for the benefits under the African Growth and Opportunity Act (AGOA), and most of its products can enter the United States market duty free. South Africa has done away with most import permits except on used products and products regulated by international treaties. It also remains committed to the simplification and continued reduction of tariffs within the WTO framework and maintains active discussions with that body and its major trading partners.

As a result of a November 1993 bilateral agreement, the Overseas Private Investment Corporation (OPIC) can assist U.S. investors in the South African market with services such as political risk insurance and loans and loan guarantees. In July 1996, the United States and South Africa signed an investment fund protocol for a $120 million OPIC fund to make equity investments in South and Southern Africa. OPIC is establishing an additional fund — the Sub-Saharan Africa Infrastructure Fund, capitalised at $350 million — to investment in infrastructure projects. The Trade and Development Agency also has been actively involved in funding feasibility studies and identifying investment opportunities in South Africa for U.S. businesses.

Despite the numerous positive economic achievements since 1994, South Africa has struggled to attract significant Foreign Direct Investment. The situation may have started to change however, with 2005 seeing the largest single FDI into South

Africa when Barclays bought a majority share in local bank Absa Group Limited. Deals between the British based Vodafone and South Africa's Vodacom have taken place in 2006.

Complicating Factors

Human Capital Flight

There has been a large degree of human capital flight from South Africa in recent years. To quote the first of the preceding references:

South Africa has lost 25% of its graduates to the United States alone. Moreover, South Africans account for 9.7% of all international medical graduates practicing in Canada. Out of all the medical graduates produced by the University of Witwatersrand in the last 35 years, more than 45% (or 2,000 physicians), have left the country. South Africa's Bureau of Statistics estimates that between 1 million and 1.6 million people in skilled, professional, and managerial occupations have emigrated since 1994 and that, for every emigrant, 10 unskilled people lose their jobs.

There are a range of causes cited for the migration of skilled South Africans. In mid 1998, the Southern African Migration Project (SAMP) undertook a study to examine and assess the range of factors that contribute to skilled South Africans' desire to leave the country:

Over two-thirds of the sample said that they had given the idea of emigration some thought while 38% said they had given it a "great deal of thought". Among the reasons cited for wishing to leave the country was the declining quality of life. Indeed, it is a common belief that the South African brain drain is heavily driven by perceptions of deteriorating quality of life since the demise of apartheid. There is general dissatisfaction with the cost of living, the level of taxation, safety and security, and the standard of public and commercial services in South Africa.

Furthermore, the government's affirmative action policy was identified as another factor influencing the emigration of skilled white South Africans. The results of the survey indicate that skilled whites are strongly opposed to this policy and the arguments advanced in support of it. Indeed, it is often the case that South African debate on the brain drain tends to show the racial contours inherent in the South African flavour of global integration. The affirmative action policy in South Africa acts to reduce the availability of work for those classified as white; a large component of the highly skilled group likely to be wealthy enough to consider emigration. Emigration of Africans with low levels of education is very low, and it is well documented that the majority of the emigrating group are white, although black professionals show no qualms in cashing in their skills abroad.

However, flight of human capital in South Africa should not be attributed solely to regional factors. For example the demand for skilled labourers in the UK, US, Canada, New Zealand, and Australia has led to active recruitment programs by those countries in South Africa. These countries accounted for 75% (by volume) of recent skilled emigration with the UK receiving approximately half of annual skilled South African emigration from 1990 to 1996. It has been suggested that the role of domestic socio-political variables may be negligible:

The association of the brain drain with socio-political change has been so widely accepted that no one – so far – has challenged the trend exhibited by the official figures. However, the magnitude of the phenomenon has quickly been put into question. Doubts arose in the mid-90s as empirical findings indicated that the departures were far higher than what the Statistics South Africa figures stated. These studies were based on embassies or removal companies' data showing that more people were leaving than the statistics mentioned. This evidence was later confirmed by a punctual statistical comparison between South African emigration data and South African registered immigration to countries such as Australia and the UK. Official agencies acknowledged the fact and recognised that their figures could only include the migrants who would declare themselves as such when leaving. However, the size of the undeclared emigrant population remained unknown. This dark side of the phenomenon had to be explored in order to bring the debate to more realistic grounds. This is where the new statistics – extracted from the receiving countries' data –

came into the picture. They showed that, even though the figures were much higher than reported by official South African data, the net loss had not begun with the political transition and was as much due to a decrease in immigration as to an increase in emigration. The concern then moved to the less emotional and more practical matter of reformulating the policy along this new perception. In this process the Ministry of Home Affairs has come under criticism. Since 1994, it has indeed followed a hard line approach towards immigration, with emphasis on controlling inflows of foreign citizens and limiting their presence and competition on the national labour market. This policy, trying to address a problem of widespread illegal immigration and consequent xenophobic trends, has been judged as inadequate regarding highly skilled people. There is now a consensus that they should rather be encouraged to come to South Africa. However, the debate continues, on the ways to make this happen. Indeed, several analysts think that the new law on immigration, coming late and with bureaucratic inertia, should be completed by more proactive dispositions, to look for skills instead of just facilitating their recruitments.

A widespread skills drain in South Africa and in the developing world in general is generally considered to be a cause for concern. While it may be the case that the economy will survive intact, the poor in South Africa undoubtedly suffer the most. The health sector has been hit particularly hard:

The report describes the exodus of health-care workers from areas of poverty and low socio-economic development, to more highly developed areas. The flows follow a hierarchy of 'wealth' and result in a global conveyor belt of health personnel moving from the bottom to the top, increasing inequity. The report describes personnel flows and migration from rural to urban areas, from public to private sectors, from lower to higher income countries within southern Africa and from African countries to industrialised countries. International migration further increases and exacerbates inequities that exist between the public and private sector and between urban and rural areas. The knowledge and skills loss from the poorer to the richer countries is considered as a form of reverse (poor to rich) subsidy. There are a variety of push and pull factors that impact on the movement of health-care workers, arising both within and beyond the health system. Factors endogenous to the health care system are low remuneration levels, work associated risks including of diseases like HIV/AIDS and TB, inadequate human resource planning with consequent unrealistic work loads, poor infrastructure and sub-optimal conditions of work. Exogenous push factors are also noted, including political insecurity, crime, taxation levels, repressive political environments and falling service standards. Movement is also influenced by pull factors, including aggressive recruitment by recipient countries, improved quality of life, study and specialisation opportunities and improved pay.

This means that emigrating medical staff do not remain to assist in the fight against HIV/AIDS, exacerbating a dire situation; South Africa has the largest population living with AIDS in the world.

Effect of HIV/AIDS

Prevalence Rate

South Africa is the country with the largest number of HIV infections in the world. The country's Department of Health estimates that 18.3% of adults (15–49 years) were living with HIV in 2006. More than half (55%) of all South Africans infected with HIV reside in the Kwa Zulu-Natal and Gauteng provinces. Rising death rates lowered life expectancy at birth from 59 years in 1990 to 49 years in 2006 for males and from 67 years in 1990 to 52.5 years in 2006 for females.

HIV prevalence among pregnant women is highest in the populous Kwa Zulu-Natal province (37%), and lowest in the Western Cape (13%), Northern Cape (16%) and Limpopo (18%) provinces. In the five other provinces (Eastern Cape, Free State, Gauteng, Mpumalanga and North West) at least 26% of women attending antenatal clinics in 2006 tested HIV-positive.

The latest HIV data collected at antenatal clinics suggest that HIV infection levels might be levelling off, with HIV prevalence in pregnant women at 30% in 2005, 29% in 2006, and 28% in 2007. The decrease in the percentage of young pregnant women (15–24 years) found to be infected with HIV also suggests a possible decline in the annual number of new infections.

Impact

HIV/AIDS has a tremendous impact on all sectors of the South Africa's economy, this includes microeconomic and macroeconomic perspectives. A study on its economic impact singles out the following effects:

- A decline in total labour supply
- A decline in labour productivity resulting from HIV/AIDS-related morbidity
- Increased production costs, prices, and a decline in aggregate demand, savings and investment
- Increased household expenditure
- Increased government expenditure.

To put a price tag to the claim of increased operation cost, Daimler-Chrysler South Africa evaluated in 2002 that every new employee infection costs the company an average of US$ 31,000.

Economic Policy

GEAR

The Government of South Africa demonstrated its commitment to open markets, privatisation and a favourable investment climate with its introduction of the Growth, Employment and Redistribution (GEAR) strategy-the neoliberal economic strategy to cover 1996-2000. Introduced by Finance Minister Trevor Manuel in June 1996, the policy set government the goals of achieving sustained annual real GDP growth of 6% or more by the year 2000 while creating 400,000 new jobs each year. The policy was meant to increase investment, especially Foreign Direct Investment, in the country to help achieve these goals.

The outcomes of the GEAR strategy have been mixed. It brought greater financial discipline and macroeconomic stability but largely failed to deliver in key areas. Formal employment continued to decline, and despite the ongoing efforts of black empowerment and signs of a fledgling black middle class and social mobility, the country's wealth remained unevenly distributed along racial lines. The desperately needed FDI also remained elusive, and consequently the ambitious economic growth targets were never realised. The policy came under stringent fire from many critics, especially when growth slumped to only 0.8% (later revised even lower to 0.5% by Statistics South Africa) in 1998.

South Africa's budgetary reforms such as the Medium-Term Expenditure Framework and the Public Finance Management Act-which aims at better reporting, auditing, and increased accountability-and the structural changes to its monetary policy framework (including inflation targeting) have, however, created transparency and predictability and are widely acclaimed. Trade liberalisation also progressed substantially since the early 1990s. Average import tariffs in South Africa, for example, declined to 14.3% in 1999 from more than 30% in 1990. These efforts, together with South Africa's implementation of its World Trade Organisation (WTO) obligations and its constructive role in launching the Doha Development Round, show South Africa's acceptance of free market principles.

One of the key pillars of the GEAR macroeconomic strategy was to reduce the fiscal deficit, which had reached over 9% of GDP during the 1993/4 fiscal year. The deficit has remained below 3% since the implementation of the reforms, greatly improving South Africa's fiscal health. The Government's 2002 budget called for a moderate increase in spending to promote faster growth and poverty alleviation.

Inflation Targeting and GDP Growth

In the February 2000 Speech, the Minister of Finance, announced a policy of inflation targeting, helping to bring consumer inflation, which had been running in the double digits for over 20 years, under control. Inflation declined from 6.9% in 1998 to less than 6.0% in 2000. The target was set to keep the South African consumer price index (CPIX) — a key indicator of inflation — between 3% and 6% average per annum. Although initially successful, the rand's rapid depreciation in late 2001 led to greater inflationary pressure and the South African Reserve Bank missed the target during the course of

2002, with inflation coming in at an average of 9.3% for the year.

From September 2003 to 2005, however, the CPIX inflation rate has remained consistently within the target range. The average annual rates of CPIX since 2001 were: 2001-6.6%, 2002-9.3%, 2003-6.8%, 2004-4.3%, 2005-4.3%.

Success in keeping inflation down allowed the Reserve Bank to reduce the *prime lending rate* — that determines the *interest rate.* During 2003 alone interest rates were cut by 550 basis points (5.5%), while between 2002 and 2006 interest rates were cut by a total 650 basis points (6.5%).

The cut in interest rates saw consumer spending rise, the construction sector boom and the sale of new vehicles reach record levels. This in turn generated much needed growth in gross domestic product (GDP). Ironically enough, GDP growth started to gather steam just as the end of the GEAR period neared. Since 1999, quarterly GDP growth has been consistently positive and annual GDP growth consistently above 2%. Between 1996 and 2004, GDP growth averaged 3.1%, rising to 4.5% (based on 2005 market prices) in 2004.

Not all economist agree with inflation targeting and the dogmatic adherence to this policy in 2006 led an successive increases in the prime lending rate that totalled 5.5%. These increases were in response to rising consumers prices, but critically consumer prices increased due to external factors (2007–2008 world food price crisis and rising oil prices). No increase in the prime lending rate could counteract these external factors and while inflation remained high the housing market and the motor manufacturing and retail sector suffered heavy losses which in turn led to heavy job losses. Even though the prime lending rate had returned to 2006 levels by mid-2009 amid the global Financial crisis of 2007–2010, it still remains high at 7%. In 2009 the Nobel Prize winning economist Joseph Stiglitz warned South Africa that inflation targeting should be a secondary concern amid the global financial crisis of 2007–2009.

Although economic growth has improved, the growth has been largely jobless, and quicker growth is still needed. The South African Government estimates that the economy must achieve growth at an average of 4.5% until 2010 and 6% thereafter to reach its goal of halving South Africa's high levels of unemployment, estimated at 26.5% (March 2005-Stats SA), by 2014.

Financial Policy

South Africa has a sophisticated financial structure with the JSE Securities Exchange, a large and active stock exchange that ranks 18th in the world in terms of total market capitalisation as of March 2009. The South African Reserve Bank (SARB) performs all central banking functions. The SARB is independent and operates in much the same way as Western central banks, influencing interest rates and controlling liquidity through its interest rates on funds provided to private sector banks. Quantitative credit controls and administrative control of deposit and lending rates have largely disappeared. South African banks adhere to the Bank of International Standards core standards.

The South African Government has taken steps to gradually reduce remaining foreign exchange controls, which apply only to South African residents. Private citizens are now allowed a one-time investment of up to R2 000 000 in offshore accounts. Since 2001, South African companies may invest up to R750 million in Africa and R500 million elsewhere. Smaller South Africa Companies can also move up to 50 Million Rand without SARB approval, allowing for swifter expansion to overseas markets. South Africa also has a strict policy of reducing its international debt and maintaining a healthy balance of trade. This has led to recent legislation promoting South African products through the Proudly South African campaign and new labelling legislation dictating all products must be labelled with their country of manufacture.

Environment

South Africa's government is deeply concerned about managing the country's rich and varied natural resources in a responsible and sustainable manner. In addition, numerous South African non-governmental organisations have emerged as a potent force in the public policy debate on the environment. In international environmental organisations, South Africa is seen as a key leader among developing countries on issues such as climate change, conservation, and biodiversity. This leading role was underscored by South Africa's selection to be the host of the World Summit on Sustainable Development in 2002. However, environmental concerns often take second tier when perceived to be a threat to business or development.

South Africa is a disproportionately large producer of carbon emissions with much of its relatively cheap electricity produced by coal-fired power stations. However, recently, due in part to UN Environmental reports and recent water restrictions and climatic fluctuations, the South African government has started formulating legislation to mitigate the effects of climate change.

Social Services

Since 1994, the government has channelled substantial resources into social programs and services, with varying degrees of success.

- Households with access to clean water: 85% in 2001, 80% in 1996
- Households using electricity for lighting: 69.7% in 2001, 57.6% in 1996
- Households in formal housing: 63.8% in 2001, 57.5% in 1996
- Households with chemical or flush toilets: 51.9% in 2001, 50.5% in 1996
- Pupil-teacher ratio: 38:1 in 2003, 43:1 in 1994
- People who have completed grade 12 schooling: 20.4% in 2001, 16.3% in 1996
- People with access to electricity: 70% in 2003, 32% in 1994
- Social grants: 6.8 million people (R34.8 billion) in 2003 and 11.8 million (October 2009).

Statistics

HDI Rank: 121st (2007) 120th (2005), 119th (2004), 111th (2003), 101st (1999), 95th (1995)

Industrial production growth rate: 5% (2004 est.), 7% (2001 est.)

Electricity:

- *production:* 221.9 TWh (2004), 213.4 TWh (2003), 206.0 TWh (2002), 196.0 TWh (2001), 195.6 TWh (2000)
- *consumption:* 204.26 TWh (2004)
- *exports:* 12.45 TWh (2004), 10.14 TWh (2003), 6.95 TWh (2002), 6.52 TWh (2001), 4.01 TWh (2000)
- *imports:* 8.03 TWh (2004), 6.74 TWh (2003), 7.87 TWh (2002), 7.25 TWh (2001), 4.72 TWh (2000).

Electricity-production by source:

- *fossil fuel:* 93.5%
- *hydroelectric:* 1.1%
- *nuclear:* 5.5%
- *other:* 0% (2001).

Total energy consumption by type:

- *Coal:* 75.4%
- *Oil:* 20.1%
- *Nuclear:* 2.8%
- *Natural gas* 1.6%
- *Hydroelectric:* 0.1%
- *other:* 0% (2004).

Agriculture-products: maize, wheat, sugarcane, fruits, vegetables; beef, poultry, mutton, wool, dairy products, essential oils;

Exports-commodities: gold, diamonds, other metals and minerals, machinery and equipment.

Imports-commodities: machinery, foodstuffs and equipment, chemicals, petroleum products, scientific instruments.

Debt-external: $25.9 billion (2004 est.)

Foreign exchange reserves: $17.618 billion (Nov 2005) $14.943 billion (Jan 2005), $6.5 billion (Oct 2003).

Exchange rates: Rand per USD (Avg Interbank rate-newest rate avg for months available).

6.16 (2006), 6.38 (2005), 6.46 (2004), 7.57 (2003)

10.5 (2002), 8.61 (2001), 6.94 (2000), 6.11 (1999)

5.53 (1998), 4.61 (1997), 4.30 (1996), 3.63 (1995)

3.55 (1994), 3.26 (1993), 2.85 (1992), 2.76 (1991)

2.58 (1990).

Weakest historical level: \$1 = R13.85 (21 December 2001)

Strongest historical level: R1 = \$1.49 (5 June 1973).

Historical annual growth in real GDP at 2005 market prices

4.5% (2004), 3.0% (2003), 3.7% (2002), 2.7% (2001)

4.2% (2000), 2.4% (1999), 0.5% (1998)

Average annual real GDP growth rate (1996–2004): 3.1%

Note: *GDP data drawn from official Stats S.A. revised statistics as released in Q3 200.*

By UN classification South Africa is a middle-income country with an abundant supply of resources, well-developed financial, legal, communications, energy, and transport sectors, a stock exchange that ranks among the top twenty in the world, and a modern infrastructure supporting an efficient distribution of goods to major urban centres throughout the entire region. South Africa is ranked 25th in the world in terms of GDP (PPP) as of 2008.

Advanced development is significantly localised around four areas: Cape Town, Port Elizabeth, Durban, and Pretoria/ Johannesburg. Beyond these four economic centres, development is marginal and poverty is still prevalent despite government efforts. Consequently the vast majority of South Africans are poor. However, key marginal areas have experienced rapid growth recently. Such areas include Mossel Bay to Plettenberg Bay; Rustenburg area; Nelspruit area; Bloemfontein; Cape West Coast; and the Kwa Zulu-Natal North Coast.

Unemployment is extremely high and income inequality is approximately equal to Brazil. During 1995–2003, the number of formal jobs decreased and informal jobs increased; overall unemployment worsened. The average South African household income decreased considerably between 1995 and 2000. As for racial inequality, Statistics South Africa reported that in 1995 the average white household earned four times as much as the average black household. In 2000 the average white household was earning six times more than the average black household. The affirmative action policies, called Black Economic Empowerment, have seen a rise in black economic wealth and an emerging black middle class. Other problems are crime, corruption, and HIV/AIDS. South Africa suffers from relatively heavy overall regulation burden compared to developed countries. State ownership and interference impose high barriers to entry in many areas. Restrictive labour regulations have contributed to the unemployment malaise.

The 1994 government inherited an economy wracked by long years of internal conflict and external sanctions. The government refrained from resorting to economic populism. Inflation was brought down, public finances were stabilised, and some foreign capital was attracted. However, growth was still subpar. At the start of 2000, then President Thabo Mbeki vowed to promote economic growth and foreign investment by relaxing restrictive labour laws, stepping up the pace of privatisation, and cutting unneeded governmental spending. His policies face strong opposition from organised labour. From 2004 onward economic growth picked up significantly; both employment and capital formation increased.

South Africa is the largest energy producer and consumer on the continent. South Africa is a popular tourist destination, and a substantial amount of revenue comes from tourism. Among the main attractions are the diverse and picturesque culture, the game reserves and the highly regarded local wines.

The South African rand (ZAR), is thc most actively traded emerging market currency in the world. It has joined an elite club of fifteen currencies, the Continuous linked settlement (CLS), where forex transactions are settled immediately, lowering the risks of transacting across time zones. The rand was the best-performing currency against the United States dollar (USD) between 2002 and 2005, according to the Bloomberg Currency Scorecard.

The volatility of the rand has affected economic activity, falling sharply during 2001 and hitting a historic low of 13.85 ZAR to the USD, raising fears of inflation, and causing the Reserve Bank to increase interest rates. The rand has since recovered, trading at 7.77 ZAR to the dollar as of February

2010. However, as exporters are put under considerable pressure from a stronger domestic currency, many call for government intervention to help soften the rand.

Refugees from poorer neighbouring countries include many immigrants from the Democratic Republic of the Congo, Mozambique, Zimbabwe, Malawi and others, representing a large portion of the informal sector. With high unemployment levels amongst poorer South Africans, xenophobia is prevalent and many people born in South Africa feel resentful of immigrants who are seen to be depriving the native population of jobs, a feeling which has been given credibility by the fact that many South African employers have employed migrants from other countries for lower pay than South African citizens, especially in the construction, tourism, agriculture and domestic service industries. Illegal immigrants are also heavily involved in informal trading. However, many immigrants to South Africa continue to live in poor conditions, and the South African immigration policy has become increasingly restrictive since 1994.

Principal international trading partners of South Africa—besides other African countries—include Germany, the United States, China, Japan, the United Kingdom and Spain. Chief exports include corn, diamonds, fruits, gold, metals and minerals, sugar, and wool. Machinery and transportation equipment make up more than one-third of the value of the country's imports. Other imports include chemicals, manufactured goods, and petroleum.

Electricity Crisis

After unsuccessful attempts by the government to encourage private construction of power generation capacity, the state-owned power supplier Eskom started experiencing deficiency in capacity in the electrical generating and reticulation infrastructure in 2007. Such lack led to inability to meet the routine demands of industry and consumers, resulting in countrywide rolling blackouts. Initially, the lack of capacity was triggered by a failure at Koeberg nuclear power station, but a general lack of capacity due to increased demand has become evident since then. The supplier has been widely criticised for failing to adequately plan for and construct sufficient electrical generating capacity, although ultimately the government has admitted that it is at fault for refusing to approve funding for investment in infrastructure.

The crisis was resolved within a few months, but the margin between national demand and available capacity is still low (particularly in peak hours), and power stations are under strain, such that another phase of rolling blackouts is probable if parts of the supply are halted for whatever reason. The government and Eskom are currently planning new power stations. The power utility plans to have 20,000 megawatts of nuclear power in its grid by 2025.

Demographics

South Africa is a nation of about 50 million people of diverse origins, cultures, languages, and religions. The last census was held in 2001 and the next will be in 2011. Statistics South Africa provided five racial categories by which people could classify themselves, the last of which, "unspecified/other" drew negligible responses, and these results were omitted. The 2009 midyear estimated figures for the other categories were Black African at 79.3%, White at 9.1%, Coloured at 9.0%, and Indian or Asian at 2.6%.

Even though the population of South Africa has increased in the past decade (primarily due to immigration), the country had an annual population growth rate of "0.051% in 2010 (CIA est.), where the birth rate is higher than the death rate but there is a net emigration rate. South Africa is home to an estimated 5 million illegal immigrants, including some 3 million Zimbabweans. A series of anti-immigrant riots occurred in South Africa beginning on 11 May 2008.

By far the major part of the population classified itself as African or black, but it is not culturally or linguistically homogeneous. Major ethnic groups include the Zulu, Xhosa, Basotho (South Sotho), Bapedi (North Sotho), Venda, Tswana, Tsonga, Swazi and Ndebele, all of which speak Bantu languages.

Some, such as the Zulu, Xhosa, Bapedi and Venda groups, are unique to South Africa. Other groups are distributed across the borders with neighbours of South Africa: The Basotho group is also the major ethnic group in Lesotho. The Tswana ethnic group constitute the majority of the population of Botswana. The Swazi ethnic group is the major ethnic group in Swaziland. The Ndebele ethnic group is also found in Matabeleland in Zimbabwe, where they are known as the Matabele. These Ndebele people are the descendants of a Zulu faction under the warrior Mzilikazi that escaped persecution from Shaka by migrating to their current territory. The Tsonga ethnic group is also found in southern Mozambique, where they are known as the Shangaan.

The white population is not ethnically homogeneous and descends from many ethnic groups: Dutch, Flemish, Portuguese, Norwegian, German, Greek, French Huguenot, English, Polish, Irish, Italian, Scottish and Welsh. Culturally and linguistically, they are divided into the Afrikaners, who speak Afrikaans, and English-speaking groups, many of whom are descended from British and Irish immigrants. Many small communities that have immigrated over the last century retain the use of other languages. There is also a substantial (though decreased) Jewish population, the majority of whom came from Lithuania at the turn of the 20th century; though others came then and later from Great Britain, the former Soviet Union and Israel. The white population is on the decrease due to a low birth rate and emigration; as a factor in their decision to emigrate, many cite the high crime rate and the affirmative action policies of the government. Since 1994, approximately 1,000,000 white South Africans have permanently emigrated.

Despite high emigration levels, a high level of non-South African white immigrants have settled in the country, in particular from countries such as Britain and Zimbabwe. For example, by 2005, an estimated 212,000 British citizens were residing in South Africa. Since 2003, the numbers of British migrants coming to South Africa has risen by 50%. An estimated 20,000 British migrants moved to South Africa in 2007. There have also been a significant number of white Zimbabwean arrivals, fleeing their home country in light of the economic and political problems currently facing the country. As well as recent arrivals, a significant number of white Zimbabweans emigrated to South Africa in the wake of independence in Zimbabwe in 1980. Some of the more nostalgic members of the community are known in popular culture as "Whenwes", because of their nostalgia for their lives in Rhodesia "when we were in Rhodesia". There have been other white immigration waves to South Africa in recent decades. In the 1970s, many Portuguese residents of African colonies such as Angola and Mozambique, came to live in South Africa after the independence of those nations. In addition, the apartheid government encouraged Eastern European immigration in the 1980s and early 1990s, particularly from Poland and Hungary.

The term "coloured" is still used for the people of mixed race descended from slaves brought in from East and Central Africa, the indigenous Khoisan who lived in the Cape at the time, Bantus, Whites (mostly the Dutch/Afrikaner and British settlers) as well as an admixture of Javanese, Malay, Indian, Malagasy and Asian blood (such as Burmese). The majority speak Afrikaans. Khoisan is a term used to describe two separate groups, physically similar: light-skinned and small in stature. The Khoikhoi, who were called *Hottentots* by the Europeans, were pastoralists and were annihilated; the San, called Bushmen by the Europeans, were hunter-gatherers. Within the Coloured community, more recent immigrants will also be found: Coloureds from the former Rhodesia (now Zimbabwe); Namibia and immigrants of mixed descent from India and Burma (Anglo-Indians/Anglo-Burmese) who were welcomed to the Cape when India and Burma received their Independence.

The major part of the South African Asian population is Indian in origin; many of them descended from indentured workers brought in the nineteenth century to work on the sugar plantations of the eastern coastal area then known as Natal. Serious riots in Durban between Indians and Zulus erupted in 1949. There is also a significant group of Chinese South Africans (approximately 100,000 individuals) and Vietnamese South Africans (approximately 50,000 individuals). In 2008, the Pretoria High Court has ruled that Chinese South Africans who arrived before 1994 are to be reclassified as Coloureds. As a result of this ruling, about 12,000–15,000 ethnically Chinese citizens who arrived before 1994, numbering 3%–5% of the total Chinese population in the country, will be able to benefit from government BEE policies. South Africa hosts a sizeable refugee and asylum seeker population. According to the *World Refugee Survey 2008*, published by the

U.S. Committee for Refugees and Immigrants, this population numbered approximately 144,700 in 2007. Groups of refugees and asylum seekers numbering over 10,000 included people from Zimbabwe (48,400), The Democratic Republic of the Congo (24,800), and Somalia (12,900). These populations mainly lived in Johannesburg, Pretoria, Durban, Cape Town, and Port Elizabeth.

Health

The spread of AIDS (acquired immuno-deficiency syndrome) is an alarming problem in South Africa with up to 31% of pregnant women found to be HIV infected in 2005 and the infection rate among adults estimated at 20%. The link between HIV, a virus spread primarily by sexual contact, and AIDS was long denied by prior president Thabo Mbeki and then health minister Manto Tshabalala-Msimang, who insisted that the many deaths in the country are due to malnutrition, and hence poverty, and not HIV. In 2007, in response to international pressure, the government made efforts to fight AIDS. In September 2008 Thabo Mbeki was ousted by the ANC and Kgalema Motlanthe was appointed for the interim. One of Mr. Motlanthe's first actions was to replace Mrs. Tshabalala-Msimang with Barbara Hogan who immediately started working to improve the Government's approach to AIDS. After the 2009 General Elections, President Jacob Zuma appointed Dr. Aaron Motsoaledi as the new minister and committed his government to increasing funding for and widening the scope of AIDS treatment. AIDS affects mainly those who are sexually active and is far more prevalent in the black population. Most deaths are people who are also economically active, resulting in many families losing their primary wage earners. This has resulted in many 'AIDS orphans' who in many cases depend on the state for care and financial support. It is estimated that there are 1,200,000 orphans in South Africa. Many elderly people also lose the support from lost younger members of their family. Roughly 5 million people are infected with the disease.

Science and Technology

Several important scientific and technological developments have originated in South Africa. The first human-to-human heart transplant was performed by cardiac surgeon Christiaan Barnard at Groote Schuur Hospital in December 1967. Max Theiler developed a vaccine against Yellow Fever, Allan McLeod Cormack pioneered x-ray Computed tomography, and Aaron Klug developed crystallographic electron microscopy techniques. These advancements were all (with the exception of that of Barnard) recognised with Nobel Prizes. Sydney Brenner won most recently, in 2002, for his pioneering work in molecular biology.

Mark Shuttleworth founded an early Internet security company Thawte, that was subsequently bought out by world-leader Veri Sign. Despite government efforts to encourage entrepreneurship in biotechnology, IT and other high technology fields, no other notable ground breaking companies have been founded in South Africa. However, it is the expressed objective of the government to transition the economy to be more reliant on high technology, based on the realisation that South Africa cannot compete with Far Eastern economies in manufacturing, nor can the republic rely on its mineral wealth in perpetuity.

South Africa has cultivated a burgeoning astronomy community. It hosts the Southern African Large Telescope, the largest optical telescope in the southern hemisphere. South Africa is currently building the Karoo Array Telescope as a pathfinder for the $20 billion Square Kilometre Array project. South Africa is a finalist, with Australia, to be the host of the SKA.

Society and Culture

It may be argued that there is no single South African culture because of its ethnic diversity. Today, the diversity in foods from many cultures is enjoyed by all and especially marketed to tourists who wish to sample the large variety of South African cuisine. In addition to food, music and dance feature prominently. South African cuisine is heavily meat-based and has spawned the distinctively South African social gathering known as a *braai*, or barbecue. South Africa has also developed into a major wine producer, with some of the best vineyards lying in valleys around Stellenbosch, Franschoek, Paarl and Barrydale. The South African black majority still has a substantial number of rural inhabitants who lead largely impoverished lives. It is among these people, however, that cultural traditions survive most strongly; as blacks have become increasingly urbanised and Westernised, aspects of traditional culture have declined. Urban blacks usually speak English or Afrikaans in addition to their native tongue. There are smaller but still significant groups of speakers of Khoisan languages who are not included in the eleven official languages, but are one of the eight other officially recognised languages. There are small groups of speakers of endangered languages, most of which are from the Khoi-San family, that receive no official status; however, some groups within South Africa are attempting to promote their use and revival.

Members of middle class, who are predominantly white but whose ranks include growing numbers of black, coloured and Indian people, have lifestyles similar in many respects to that of people found in Western Europe, North America and Australasia. Members of the middle class often study and work abroad for greater exposure to the markets of the world.

Asians, predominantly of Indian origin, preserve their own cultural heritage, languages and religious beliefs, being either Christian, Hindu or Sunni Muslim and speaking English, with Indian languages like Hindi, Telugu, Tamil or Gujarati being spoken less frequently, but the majority of Indians being able to understand their mother tongue. The first Indians arrived on the famous Truro ship as indentured labourers in Natal to work the Sugar Cane Fields. There is a much smaller Chinese community in South Africa, although its numbers have increased due to immigration from Republic of China (Taiwan).

South Africa has also had a large influence in the Scouting movement, with many Scouting traditions and ceremonies coming from the experiences of Robert Baden-Powell (the founder of Scouting) during his time in South Africa as a military officer in the 1890s. The South African Scout Association was one of the first youth organisations to open its doors to youth and adults of all races in South Africa. This happened on 2 July 1977 at a conference known as *Quo Vadis*.

Music

There is great diversity in music from South Africa. Many black musicians who sang in Afrikaans or English during apartheid have since begun to sing in traditional African languages, and have developed a unique style called Kwaito. Of note is Brenda Fassie, who launched to fame with her song "Weekend Special", which was sung in English. More famous traditional musicians include Ladysmith Black Mambazo, while the Soweto String Quartet performs classic music with an African flavour. White and Coloured South African singers are historically influenced by European musical styles. South Africa has produced world-famous jazz musicians, notably Hugh Masekela, Jonas Gwangwa, Abdullah Ibrahim, Miriam Makeba, Jonathan Butler, Chris McGregor, and Sathima Bea Benjamin. Afrikaans music covers multiple genres, such as the contemporary Steve Hofmeyr and the punk rock band Fokofpolisiekar. Crossover artists such as Verity (internationally recognised for innovation in the music industry) and Johnny Clegg and his bands Juluka and Savuka have enjoyed various success underground, publicly, and abroad.

The South African music scene includes Kwaito, a new music genre that had developed in the mid 80s and has since developed to become the most popular social economical form of representation among the populous. Though some may argue that the political aspects of Kwaito has since diminished after Apartheid, and the relative interest in politics has become a minor aspect of daily life. Some argue that in a sense, Kwaito is in fact a political force that shows activism in its apolitical actions. Today, major corporations like Sony, BMG, and EMI have appeared on the South African scene to produce and distribute Kwaito music. Due to its overwhelming popularity, as well as the general influence of DJs, who are among the top 5 most influential types of people within the country, Kwaito has taken over radio, television, and magazines.

Religion

According to the 2001 national census, Christians accounted for 79.7% of the population. This includes Zion Christian (11.1%), Pentecostal (Charismatic) (8.2%), Roman Catholic (7.1%), Methodist (6.8%), Dutch Reformed (6.7%), Anglican (3.8%); members of other Christian churches accounted for another 36% of the population. Muslims accounted for 1.5% of the population, Hindus about 1.3%, and Judaism 0.2%. 15.1% had no religious affiliation, 2.3% were other and 1.4% were unspecified.

African Indigenous Churches were the largest of the Christian groups. It was believed that many of these persons who claimed no affiliation with any organised religion adhered to traditional indigenous religions. Many peoples have syncretic religious practices combining Christian and indigenous influences.

There is no evidence that Islam was known to the Zulu, Swazi, or Xhosa of the east coast prior to the colonial era. Many South African Muslims are described as Coloureds, notably in the Western Cape, especially those whose ancestors came as slaves from the Indonesian archipelago (the Cape Malays). Others are described as Indians, notably in Kwa Zulu-Natal, including those whose ancestors came as traders from South Asia; they have been joined by others from other parts of Africa as well as white or black South African converts. South African Muslims claim that their faith is the fastest-growing religion of conversion in the country.

The Hindu population was primarily established during British colonial period, but later waves of immigration from India have also contributed to it. Most Hindus are ethnically South Asian but there are many who come from mixed racial stock, and some are converts with the efforts of Hindu missionaries such as ISKCON.

Other minority religions in South Africa are Sikhism, Jainism and Bahai Faith.

Languages

South Africa has eleven official languages: Afrikaans, English, Ndebele, Northern Sotho, Sotho, Swazi, Tswana, Tsonga, Venda, Xhosa and Zulu. In this regard it is third only to Bolivia and India in number. While all the languages are formally equal, some languages are spoken more than others. According to the 2001 National Census, the three most spoken first home languages are Zulu (23.8%), Xhosa (17.6%) and Afrikaans (13.3%). Despite the fact that English is recognised as the language of commerce and science, it was spoken by only 8.2% of South Africans at home in 2001, an even lower percentage than in 1996 (8.6%).

The country also recognises eight unofficial languages: Fanagalo, Khoe, Lobedu, Nama, Northern Ndebele, Phuthi, San and South African Sign Language. These unofficial languages may be used in certain official uses in limited areas where it has been determined that these languages are prevalent. Nevertheless, their populations are not such that they require nationwide recognition.

Many of the "unofficial languages" of the San and Khoikhoi people contain regional dialects stretching northwards into Namibia and Botswana, and elsewhere. These people, who are a physically distinct population from other Africans, have their own cultural identity based on their hunter-gatherer societies. They have been marginalised to a great extent, and many of their languages are in danger of becoming extinct.

Many white South Africans also speak other European languages, such as Portuguese (also spoken by black Angolans and Mozambicans), German, and Greek, while some Asians and Indians in South Africa speak South Asian languages, such as Tamil, Hindi, Gujarati, Urdu and Telugu. French is still widely spoken by French South Africans especially in places like Franschhoek, where many South Africans are of French origin. South African French is spoken by fewer than 10,000 individuals. Congolese French is also spoken in South Africa by migrants.

Sports

South Africa's most popular sports are soccer, rugby union and cricket. Other sports with significant support are swimming, athletics, golf, boxing, tennis and netball. Although soccer commands the greatest following among the youth, other sports like basketball, surfing and skateboarding are increasingly popular. Famous boxing personalities include Baby Jake Jacob Matlala, Vuyani Bungu, Welcome Ncita, Dingaan Thobela, Gerrie Coetzee and Brian Mitchell. Soccer players who have played for major foreign clubs include Lucas Radebe and Philemon Masinga (both formerly of Leeds United), Quinton Fortune (Atletico Madrid and Manchester United), Benni McCarthy (Ajax Amsterdam, F.C. Porto, Blackburn Rovers and West Ham United), Aaron Mokoena (Ajax Amsterdam, Blackburn Rovers and Portsmouth), Delron Buckley (Borussia

Dortmund) and Steven Pienaar (Ajax Amsterdam and Everton). South Africa produced Formula One motor racing's 1979 world champion Jody Scheckter. Famous current cricket players include Herschelle Gibbs, Graeme Smith, Jacques Kallis, JP Duminy, etc. Most of them also participate in the Indian Premier League.

South Africa has also produced numerous world class rugby players, including Francois Pienaar, Joost van der Westhuizen, Danie Craven, Frik du Preez, Naas Botha and Bryan Habana. South Africa hosted and won the 1995 Rugby World Cup at their first attempt and won the 2007 Rugby World Cup in France. South Africa was only allowed to participate from 1995 since the end of Apartheid.

It followed the 1995 Rugby World Cup by hosting the 1996 African Cup of Nations, with the national team, 'Bafana Bafana,' going on to win the tournament. It also hosted the 2003 Cricket World Cup and the 2007 World Twenty20 Championship which were a great success. South Africa is the host nation for the 2010 FIFA World Cup, which is the first time the tournament is held on the African continent.

In 2004, the swimming team of Roland Schoeman, Lyndon Ferns, Darian Townsend and Ryk Neethling won the gold medal at the 2004 Olympic Games in Athens, simultaneously breaking the world record in the 4x100 freestyle relay. Penny Heyns won Olympic Gold in the 1996 Atlanta Olympic Games.

In golf, Gary Player is generally regarded as one of the greatest golfers of all time, having won the Career Grand Slam, one of five golfers to have done so. Other South African golfers to have won major tournaments include Bobby Locke, Ernie Els, Retief Goosen and Trevor Immelman.

Education

Primary schools span the first seven years of schooling. In the age of Apartheid, schools for blacks were subject to discrimination through inadequate funding and a separate syllabus called 'Bantu Education' which was only designed to give them sufficient skills to work as labourers. South Africa has numerous universities. Instruction can take place in Afrikaans as well. Public expenditure on education was at 5.4 % of the 2002-05 GDP.

Social Issues

According to a survey for the period 1998–2000 compiled by the United Nations, South Africa was ranked second for murder and first for assaults and rapes per capita. Official statistics show that 52 people are murdered every day in South Africa. The reported number of rapes per year is 55,000, and it is estimated that 500,000 rapes are committed annually in South Africa. Total crime per capita is 10th out of the 60 countries in the data set.

Rape is a common problem in South Africa, in a 2009 survey one in four South African men admitted to raping someone. One in three of the 4,000 women questioned by the Community of Information, Empowerment and Transparency said they had been raped in the past year. South Africa has some of the highest incidences of child and baby rape in the world. In a related survey conducted among 1,500 schoolchildren in the Soweto township, a quarter of all the boys interviewed said that 'jackrolling', a term for gang rape, was fun.

Middle-class South Africans seek security in gated communities. Many emigrants from South Africa also state that crime was a big motivator for them to leave. Crime against the farming community has continued to be a major problem.

Along with many African nations, South Africa has been experiencing a "brain drain" in the past 20 years. This is believed to be potentially damaging for the regional economy, and is almost certainly detrimental for the well-being of the majority of people reliant on the health-care infrastructure, given the HIV/AIDS epidemic. The skills drain in South Africa tends to demonstrate racial contours (naturally given the skills distribution legacy of South Africa) and has thus resulted in large white South African communities abroad.

In May 2008 societal hostility to African migrants exploded in a series of pogroms that left up to 100 people dead and 100,000 displaced.

Bibliography

Belden C.: *Landscapes of the Sacred: Geography and Narrative in American Spirituality*, New York: Paulist Press, 1988.

Berry, J.K. : *Beyond Mapping, Concepts, Algorithms and Issues in GIS*, Fort Collins, CO: GIS World Books, 1993.

Bolstad, P. : *GIS Fundamentals, A first text on Geographic Information Systems, Second Edition*, White Bear Lake, MN, Eider Press, 2005.

Cronon, William: *Changes in the Land: Indians, Colonists, and the Ecology of New England.* New York: Hill and Wang, 1983.

Davis, Mike: *Ecology of Fear: Los Angeles and the Imagination of Disaster*. New York: Vintage, 1999.

Elangovan K. : *GIS, Fundamentals, Applications and Population Geography*, New India Publishing Agency, New Delhi 2006.

Harvey, David: *Explanation in Geography.* London: Edward Arnold, 1969.

Harvey, Francis : *A Primer of GIS, Fundamental Geographic and Cartographic Concepts,* The Guilford Press, 2008.

Jackson, Peter: *Maps of Meaning: An Introduction to Cultural Geography.* London and Boston: Unwin Hyman, 1989.

James S. Duncan: *Writing Worlds: Discourse, Text, and Metaphor in the Representation of Landscape.* London and New York: Routledge, 1992.

James, T.: *The Best Poor Man's Country: A Geographical Study of Early Southeastern Pennsylvania*. Baltimore: Johns Hopkins Press, 1972.

King, Anthony D.: *Buildings and Society: Essays on the Social Development of the Built Enviroment*. London: Routledge & Kegan Paul, 1980.

Lewis, Peirce. *New Orleans: The Making of an Urban Landscape.* Cambridge, MA: Ballinger Publishing Co., 1976. Annotation

Liam Kennedy: *Urban Space and Representation.* Sterling, VA: Pluto Press, 2000.

Lung, Y., and van Tulder, R. : *Cars, Carriers of Regionalism*, Palgrave Macmillan, New York, 2004.

McDonnell, R.A. : *Principles of Geographical Information Systems*, Oxford University Press, Oxford, 1998.

Michael P.: *The Making of the American Landscape.* Boston: Unwin Hyman, 1990.

Michael, H.: *City Form and Natural Process: Towards a New Urban Vernacular.* London: Croom Helm, 1984.

Peter, G.: *The Slow Plague: A Geography of the AIDS Pandemic.* Oxford and Cambridge, MA: Blackwell Publishers, 1993.

Ray Allen: *The Far Western Frontier, 1830-1860.* The New American Nation Series. New York: Harper, 1956.

Richard SCASE : *Managing Creativity,* Open University Press, 2000.

Stephen, D.: *Fields of Vision: Landscape Imagery and National Identity in England and the United States.* Cambridge: Polity, 1993.

Tulder, R. : *Cars, Carriers of Regionalism*, Palgrave Macmillan, New York, 2004.

Victor, A. : *Geography of Art and Culture*, Contributions to Economic Analysis, Elsevier, 2004.

Watkins, Eric: *The Middle Eastern Environment*, London, The British Society for Middle Eastern Studie, 1995.

William G. BOWEN : *Performing Arts, The Economic Dilemma,* Thousand Oaks, SAGE, London, 2002.

Worboys, Michael, and Matt Duckham. : *GIS, a Computing Perspective*, Boca Raton, CRC Press, 2004.

Index